T0245263

CAMBRIDGE LIBRARY COLLECTION

Books of enduring scholarly value

Earth Sciences

In the nineteenth century, geology emerged as a distinct academic discipline. It pointed the way towards the theory of evolution, as scientists including Gideon Mantell, Adam Sedgwick, Charles Lyell and Roderick Murchison began to use the evidence of minerals, rock formations and fossils to demonstrate that the earth was older by millions of years than the conventional, Bible-based wisdom had supposed. They argued convincingly that the climate, flora and fauna of the distant past could be deduced from geological evidence. Volcanic activity, the formation of mountains, and the action of glaciers and rivers, tides and ocean currents also became better understood. This series includes landmark publications by pioneers of the modern earth sciences, who advanced the scientific understanding of our planet and the processes by which it is constantly re-shaped.

Hours of Exercise in the Alps

John Tyndall (1820–93) was a prominent physicist, particularly noted for his studies of thermal radiation and the atmosphere. He was a prolific writer and lecturer, who was able to bring experimental physics to a wide audience. While researching his 1860 work, *The Glaciers of the Alps*, he became a proficient climber, and this work, first published in 1871, combines climbing expeditions in Switzerland with comments on glaciation and geology. It was extremely popular, with a second edition in the same year, and German and American editions in 1872. He was one of a group of noted Alpinists of the period, making the first ascent of the Weisshorn in Switzerland and finally conquering the Matterhorn in 1868, three years after its first ascent. This account of Victorian climbing expeditions makes fascinating reading, and shows the length an experimental scientist was prepared to go in search of knowledge.

Cambridge University Press has long been a pioneer in the reissuing of out-of-print titles from its own backlist, producing digital reprints of books that are still sought after by scholars and students but could not be reprinted economically using traditional technology. The Cambridge Library Collection extends this activity to a wider range of books which are still of importance to researchers and professionals, either for the source material they contain, or as landmarks in the history of their academic discipline.

Drawing from the world-renowned collections in the Cambridge University Library, and guided by the advice of experts in each subject area, Cambridge University Press is using state-of-the-art scanning machines in its own Printing House to capture the content of each book selected for inclusion. The files are processed to give a consistently clear, crisp image, and the books finished to the high quality standard for which the Press is recognised around the world. The latest print-on-demand technology ensures that the books will remain available indefinitely, and that orders for single or multiple copies can quickly be supplied.

The Cambridge Library Collection will bring back to life books of enduring scholarly value (including out-of-copyright works originally issued by other publishers) across a wide range of disciplines in the humanities and social sciences and in science and technology.

Hours of Exercise in the Alps

JOHN TYNDALL

CAMBRIDGE
UNIVERSITY PRESS

CAMBRIDGE UNIVERSITY PRESS

Cambridge, New York, Melbourne, Madrid, Cape Town,
Singapore, São Paolo, Delhi, Tokyo, Mexico City

Published in the United States of America by Cambridge University Press, New York

www.cambridge.org
Information on this title: www.cambridge.org/9781108037822

© in this compilation Cambridge University Press 2011

This edition first published 1871
This digitally printed version 2011

ISBN 978-1-108-03782-2 Paperback

HOURS OF EXERCISE.

LONDON: PRINTED BY
SPOTTISWOODE AND CO., NEW-STREET SQUARE
AND PARLIAMENT STREET

THE JUNGFRAU FROM INTERLAKEN.

HOURS OF EXERCISE

IN

THE ALPS.

BY

JOHN TYNDALL, LL.D. F.R.S.

SECOND EDITION.

LONDON:

LONGMANS, GREEN, AND CO.

1871.

PREFACE

TO

THE SECOND EDITION.

——◆——

In 1862, at the instance of three of his friends, the
celebrated guide, Johann Joseph Bennen, came to
England; and during his visit Mr. F. F. Tuckett
had him photographed. The portrait then taken is
the basis of the capital sketch which faces page 201
of the present volume.

The engraving was executed by Mr. Whymper for
his beautifully illustrated Alpine book recently published,* and both Mr. W. Longman and myself are
much indebted to the artist for his civility in permitting the publishers of this work to make use of
his sketch.

But to the occasional blending of sweetness and
earnestness in Bennen's countenance hardly any
such portrait could do justice. It was the passing
of a gleam which threw light and tenderness into
the resolute features, but which could not be caught
by the photographer's art.

* A brief reference to Mr. Whymper's book occurs at pp. 166-67
of this volume.

As already stated in another place, the picture
from which the engraving of the Weisshorn, facing
page 91, is taken was lent to me in the most
obliging manner by Mr. Wm. Mathews, jun. The
ridge along which we ascended is to the right of a
person looking at the sketch. For the drawing
from which the engraving of the Matterhorn, facing
page 117, is taken, I have to thank my friend Mr.
E. W. Cooke, R.A.

The immediate cause of the appearance of these
' Hours of Exercise' as well as of the more serious
' Fragments' contained in another volume recently
published, was the announcement that if the work
of selection, revision, and arrangement were not executed by me, it would be executed by others without
my authority. The publication of Huxley's admirable ' Lay Sermons' created a demand for my scattered essays also, and, once committed to the task of
collecting, I thought it better to finish the work by
publishing these Alpine papers at the same time.

It is a satisfaction to me to know that, slight as
they are, they have given pleasure to persons of
gravity and good sense. The ' Fragments,' I may
say, are now being translated into German under
the supervision of Helmholtz, and the ' Hours of
Exercise' under that of Wiedemann.

July 1871.

PREFACE

TO

THE FIRST EDITION.

A SHORT TIME AGO I published a book of 'Fragments,' which might have been called 'Hours of
Exercise in the Attic and the Laboratory'; while
this one bears the title of 'Hours of Exercise in
the Alps.' The two volumes supplement each other,
and, taken together, illustrate the mode in which
a lover of natural knowledge and of natural scenery
chooses to spend his life.

Much as I enjoy the work, I do not think that I
could have filled my days and hours in the Alps with
clambering alone. The climbing in many cases was
the peg on which a thousand other 'exercises' were
hung. The present volume, however, is for the
most part a record of bodily action, written partly to
preserve to myself the memory of strong and joyous
hours, and partly for the pleasure of those who find
exhilaration in descriptions associated with mountain life.

The papers, written during the last ten years, are

printed in the order of the incidents to which they
relate ; and, to render the history more complete,
I have, with the permission of their authors, introduced nearly the whole of two articles by Mr.
Vaughan Hawkins and Mr. Philip Gossett. The
former describes the first assault ever made upon the
Matterhorn, the latter an expedition which ended in
the death of a renowned and beloved guide.

The 'Glaciers of the Alps' being out of print,
I can no longer refer to it. Towards the end of the
volume, therefore, I have thrown together a few
'Notes and Comments' which may be useful to
those who desire to possess some knowledge of the
phenomena of the ice-world, and of the properties
of ice itself. To these are added one or two minor
articles, which relate more or less to our British
hills and lakes: the volume is closed by an account
of a recent voyage to Oran.

I refrain from giving advice, further than to say
that the perils of wandering in the High Alps are
terribly real, and are only to be met by knowledge,
caution, skill, and strength. 'For rashness, ignorance, or carelessness the mountains leave no margin ;
and to rashness, ignorance, or carelessness three-
fourths of the catastrophes which shock us are
to be traced.' Those who wish to know something
of the precautions to be taken upon the peaks
and glaciers cannot do better than consult the

excellent little volume lately published by Leslie Stephen, where, under the head of 'Dangers of Mountaineering,' this question is discussed.

I would willingly have published this volume without illustrations, and should the reader like those here introduced—two of which were published ten years ago, and the remainder recently executed under the able superintendence of Mr. Whymper— he will have to ascribe his gratification to the initiative of Mr. William Longman, not to me.

I have sometimes tried to trace the genesis of the interest which I take in fine scenery. It cannot be wholly due to my own early associations; for as a boy I loved nature, and hence, to account for that love, I must fall back upon something earlier than my own birth. The forgotten associations of a far-gone ancestry are probably the most potent elements in the feeling. With characteristic pene- tration, Mr. Herbert Spencer has written of the growth of our appreciation of natural scenery with growing years. But to the associations of the indi- vidual himself he adds 'certain deeper, but now vague, combinations of states, that were organised in the race during barbarous times, when its pleasurable activities were among the mountains, woods, and waters. Out of these excitations,' he adds, 'some of them actual, but most of them nascent, is composed the emotion which a fine landscape produces in us.' I think this an exceedingly likely proximate hypothesis, and hence infer that those 'vague and deep combinations organised in barbarous times,' not to go further back, have come down with considerable force to me. Adding to these inherited feelings the pleasurable present exercise of Mr. Bain's 'muscular sense,' I obtain a somewhat intelligible, though, doubtless, still secondary theory of my delight in the mountains.

The name of a friend whom I taught in his boyhood to handle a theodolite and lay a chain, and who afterwards turned his knowledge to account on the glaciers of the Alps, occurs frequently in the following pages. Of the firmness of a friendship, uninterrupted for an hour, and only strengthened by the weathering of six-and-twenty years, he needs no assurance. Still, for the pleasure it gives myself, I connect this volume with the name of THOMAS ARCHER HIRST.

J. TYNDALL.

May 1871.

CONTENTS.

NOTES ON ICE AND GLACIERS, ETC.

ILLUSTRATIONS.

'Nature, thou earliest gospel of the wise,
   Thou never-silent hymner unto God;
Thou angel-ladder lost amidst the skies,
   Though at the foot we dream upon the sod;
To thee the priesthood of the lyre belong—
They hear religion and reply in song.

'If he hath held thy worship undefiled
   Through all the sins and sorrows of his youth,
Let the man echo what he heard as child
   From the far hill-tops of melodious Truth,
Leaving in troubled hearts some lingering tone
Sweet with the solace thou hast given his own.'

                              LORD LYTTON's *King Arthur*.

                     'The brain,
That forages all climes to hue its cells,
Will not distil the juices it has sucked
To the sweet substance of pellucid thought,
Except for him who hath the secret learned
To mix his blood with sunshine, and to take
The winds into his pulses.'

                              JAMES RUSSELL LOWELL.

HOURS OF EXERCISE

IN

THE ALPS.

————◦————

I.

THE LAUWINEN-THOR.

In June 1860 I completed 'The Glaciers of the Alps,'
which constituted but a fraction of the work executed during the previous autumn and spring.
These labours and other matters had wearied and
weakened me beyond measure, and to gain a little
strength I went to Killarney. The trip was bene-
ficial, but not of permanent benefit. The air of
those most lovely lakes was too moist and warm for
my temperament, and I longed for that keener air
which derives its tone from contact with the Alpine
snows. In 1859 I had bidden the Alps farewell,
purposing in future to steep my thoughts in the
tranquillity of English valleys, and confine my
mountain work to occasional excursions in the
Scotch Highlands, or amid the Welsh and Cumbrian
hills. But in my weariness the mere thought of

B

the snow-peaks and glaciers was an exhilaration; and to the Alps, therefore, I resolved once more to go. I wrote to my former guide, Christian Lauener, desiring him to meet me at Thun on Saturday the 4th of August; and on my way thither I fortunately fell in with Mr. Vaughan Hawkins. He told me of his plans and wishes, which embraced an attack upon the Matterhorn. Infected by his ardour, I gladly closed with the proposition that we should climb together for a time.

Lauener was not to be found at Thun, but in driving from Neuhaus to Interlaken a chaise met us, and swiftly passed; within it I could discern the brown visage of my guide. We pulled up and shouted, the other vehicle stopped, Lauener leaped from it, and came bounding towards me with admirable energy, through the deep and splashing mud. 'Gott! wie der Kerl springt!' was the admiring exclamation of my coachman. Lauener is more than six feet high, and mainly a mass of bone; his legs are out of proportion, longer than his trunk; and he wears a short-tail coat, which augments the apparent discrepancy. Those massive levers were now plied with extraordinary vigour to project his body through space; and it was gratifying to be thus assured that the man was in first-rate condition, and fully up to the hardest work.

On Sunday the 5th of August, for the sake of a

little training, I ascended the Faulhorn alone. The
morning was splendid, but as the day advanced
heavy cloud-wreaths swathed the heights. They
attained a maximum about two P.M., and afterwards
the overladen air cleared itself by intermittent jerks
—revealing at times the blue of heaven and the
peaks of the mountains; then closing up again, and
hiding in their dismal folds the very posts which stood
at a distance of ten paces from the hotel door. The
effects soon became exceedingly striking, the mutations were so quick and so forcibly antithetical. I
lay down upon a seat, and watched the intermittent
extinction and generation of the clouds, and the alternate appearance and disappearance of the mountains. More and more the sun swept off the swelter-
ing haze, and the blue sky bent over me in domes of
ampler span. At four P.M. no trace of cloud was
visible, and a panorama of the Oberland, such as I had
no idea that the Faulhorn could command, unfolded
itself. There was the grand barrier which separated
us from the Valais; there were the Jungfrau, Monk
and Eiger, the Finsteraarhorn, the Schreckhorn, and
the Wetterhorn, lifting their snowy and cloudless
crests to heaven, and all so sharp and wildly precipitous that the bare thought of standing on any one of
them made me shudder. London was still in my
brain, and the vice of Primrose Hill in my muscles.

I disliked the ascent of the Faulhorn exceedingly,

and the monotonous pony track which led to the top of it. Once, indeed, I deviated from the road out of pure disgust, and, taking a jumping torrent for my guide and colloquist, was led astray. I now resolved to return to Grindelwald by another route. My host at first threw cold water on the notion, but he afterwards relaxed and admitted that the village might be attained by a.more direct way than the ordinary one. He pointed to some rocks, eminences, and trees, which were to serve as landmarks; and stretching his arm in the direction of Grindelwald, I took the bearing of the place, and scampered over slopes of snow to the sunny Alp beyond them. To my left was a mountain stream making soft music by the explosion of its bubbles. I was once tempted aside to climb a rounded eminence, where I lay for an hour watching the augmenting glory of the mountains. The scene at hand was perfectly pastoral ; green pastures, dotted with chalets, and covered with cows, which filled the air with the incessant tinkle of their bells. Beyond was the majestic architecture of the Alps, with its capitals and western bastions flushed with the warm light of the lowering sun.

I mightily enjoyed the hour. There was health in the air and hope in the mountains, and with the consciousness of augmenting vigour I quitted my station, and galloped down the Alp. I was soon

amid the pinewoods which overhang the valley of
Grindelwald, with no guidance save the slope of the
mountain, which, at times, was quite precipitous;
but the roots of the pines grasping the rocks afforded
hand and foot such hold as to render the steepest
places the pleasantest of all. I often emerged from
the gloom of the trees upon lovely bits of pasture—
bright emerald gems set in the bosom of the woods.
It appeared to me surprising that nobody had constructed a resting-place on this fine slope. With a
fraction of the time necessary to reach the top of the
Faulhorn, a position might be secured from which the
prospect would vie in point of grandeur with almost
any in the Alps; while the ascent from Grindelwald,
amid the shade of the festooned trees, would itself
be delightful.

Hawkins, who had halted for a day at Thun, had
arrived; our guide had prepared a number of stakes,
and on Monday morning we mounted our theodolite
and proceeded to the Lower Glacier. With some
difficulty we established the instrument upon a site
whence the glacier could be seen from edge to edge;
and across it was fixed in a straight line a series of
twelve stakes. We afterwards ascended the glacier
till we touched the avalanche-débris of the Heisse
Platte. We wandered amid the moulins and crevasses until evening approached, and thus gradually
prepared our muscles for more arduous work. On

Tuesday a sleety rain filled the entire air, and the glacier was so laden with fog that there was no possibility of our being able to see across it. On Wednesday, happily, the weather brightened, and we executed our measurements; finding, as in all other cases, that the glacier was retarded by its bounding walls, its motion varying from a minimum of thirteen and a half inches to a maximum of twenty-two inches a day. To Mr. Hawkins I am indebted both for the fixing of the stakes and the reduction of the measurements to their diurnal rate.

Previous to leaving England I had agreed to join a party of friends at the Æggischhorn, on Thursday the 9th of August. My plan was, first to measure the motion of the Grindelwald glacier, and afterwards to cross the mountain-wall which separates the canton of Berne from that of Valais, so as to pass from Lauterbrunnen to the Æggischhorn in a single day. How this formidable barrier was to be crossed was a problem, but I did not doubt being able to get over it somehow. On mentioning my wish to Lauener, he agreed to try, and proposed attacking it through the Roththal. In company with his brother Ulrich, he had already spent some time in the Roththal, seeking to scale the Jungfrau from that side. Hawkins had previously, I believe, entertained the thought of assailing the same barrier at the very same place. Having completed our

measurements on the Wednesday, we descended to
Grindelwald and discharged our bill. We desired to
obtain the services of Christian Kaufmann, a guide
well acquainted with both the Wetterhorn and the
Jungfrau; but on learning our intentions he expressed fears regarding his lungs, and recommended
to us his brother, a powerful young man, who had
also undergone the discipline of the Wetterhorn.
Him we accordingly engaged. We arranged with
the landlord of the Bear to have the main mass of
our luggage sent to the Æggischhorn by a more
easy route. I was loth to part with the theodolite,
but Lauener at first grumbled hard against taking
it. It was proposed, however, to confine his load to
the head of the instrument, while Kaufmann should
carry the legs, and I should bear my own knapsack.
He yielded. Ulrich Lauener was at Grindelwald
when we started for Lauterbrunnen, and on bidding
us good-bye he remarked that we were going to
attempt an impossibility. He had examined the
place which we proposed to assail, and emphatically
affirmed that it could not be surmounted. We were
both a little chagrined by this gratuitous announcement, and answered him somewhat warmly; for we
knew the moral, or rather immoral, effect of such an
opinion upon the spirits of our men.

The weather became more serene as we approached
Lauterbrunnen. We had a brief evening stroll, but
retired to bed before day had quite forsaken the

mountains. At two A.M. the candle of Lauener gleamed into our bedrooms, and he pronounced the weather fair. We got up at once, dressed, despatched our hasty breakfast, strapped our things into the smallest possible volume, and between three and four A.M. were on our way. The hidden sun crimsoned faintly the eastern sky, but the valleys were all in peaceful shadow. To our right the Staubbach dangled its hazy veil, while other *Bachs* of minor note also hung from the beetling rocks, but fell to earth too lightly to produce the faintest murmur. After an hour's march we deviated to the left, and wound upward through the woods which here cover the slope of the hill.

The dawn cheerfully unlocked the recesses of the mountains, and we soon quitted the gloom of the woods for the bright green Alp. This we breasted, regardless of the path, until we reached the chalets of the Roththal. We did not yet see the particular staircase up which Lauener proposed to lead us, but we inspected minutely the battlements to our right, marking places for future attack in case our present attempt should not be successful. The elastic grass disappeared, and we passed over rough crag and shingle alternately. We reached the base of a ridge of *débris*, and mounted it. At our right was the glacier of the Roththal, along the lateral moraine of which our route lay.

Just as we touched the snow a spring bubbled from the rocks at our left, spurting its water over stalagmites of ice. We turned towards it, and had each a refreshing draught. Lauener pointed out to us the remains of the hut erected by him and his brother when they attempted the Jungfrau, and from which they were driven by adverse weather. We entered an amphitheatre, grand and beautiful this splendid morning, but doubtless in times of tempest a fit residence for the devils whom popular belief has banished to its crags. The snow for a space was as level as a prairie, but in front of us rose the mighty bulwarks which separated us from the neighbouring canton. To our right were the crags of the Breithorn, to our left the buttresses of the Jungfrau, while between both was an indentation in the mountain-wall, on which all eyes were fixed. From it downwards hung a thread of snow, which was to be our leading-string to the top.

Though very steep, the aspect of the place was by no means terrible: comparing with it my memory of other gulleys in the Chamouni mountains, I imagined that three hours would place us at the top. We not only expected an easy conquest of the barrier, but it was proposed that on reaching the top we should turn to the left, and walk straight to the summit of the Jungfrau. Lauener was hopeful, but not sanguine. We were soon at the foot of

the barrier, clambering over mounds of snow. Huge
consolidated lumps emerged from the general mass;
the snow was evidently that of avalanches which had
been shot down the couloir, kneading themselves
into vast balls, and piling themselves in heaps upon
the plain. The gradient steepened, the snow was
hard, and the axe was invoked. Straight up the
couloir seemed the most promising route, and we
pursued it for an hour, the impression gradually
gaining ground that the work would prove heavier
than we had anticipated.

We then turned our eyes on the rocks to our
right, which seemed practicable, though very steep;
we swerved towards them, and worked laboriously
upwards for three-quarters of an hour. Mr. Hawkins
and the two guides then turned to the left, and
regained the snow, leaving me among the crags.
They had steps to cut, while I had none, and, consequently, I got rapidly above them. The work becomes ever harder, and rest is unattainable, for there
is no resting-place. At every brow I pause; legs
and breast are laid against the rough rock, so as to
lessen by their friction the strain upon the arms,
which are stretched to grasp some protuberance
above. Thus I rest, and thus I learn that three
days' training is not sufficient to dislodge London
from one's lungs. Meanwhile my companions are
mounting monotonously along the snow. Lauener

ASCENT OF THE LAUWINEN THOR.

looks up at me at intervals, and I can clearly mark the
expression of his countenance ; it is quite spiritless,
while that of his companion bears the print of absolute
dismay. Three hours have passed and the summit
is not sensibly nearer. The men halt and converse
together. Lauener at length calls out to me, ' I
think it is impossible.' The effect of Ulrich's prediction appears to be cropping out; we expostulate,
however, and they move on. After some time they
halt again, and reiterate their opinion that the thing
cannot be done. They direct our attention to the
top of the barrier; light clouds scud swiftly over it,
and snow-dust is scattered in the air. There is
storm on the heights, which our guides affirm has
turned the day against us. I cast about in my mind
to meet the difficulty, and enquire whether we might
not send one of them back with the theodolite, and
thus so lighten our burdens as to be able to proceed.
Kaufmann volunteers to take back the theodolite ;
but this does not seem to please Lauener. There
is a pause and hesitation. I remonstrate, while
Hawkins calls out 'Forward !' Lauener once more
doggedly strikes his axe into the snow, and resumes
the ascent.

I continued among the rocks, though with less
and less confidence in the wisdom of my choice.
My knapsack annoyed me excessively; the straps
frayed my shoulders, and tied up my muscles.

Once or twice I had to get round a protruding face
of rock, and then found my bonds very grievous.
At length I came to a peculiar piece of cliff, near
the base of which was a sharp ridge of snow, and
at a height of about five feet above it the rock
bulged out, so that a stone dropped from its protuberance would fall beyond the ridge. I had to
work along the snow cautiously, squatting down so
as to prevent the rock from pushing me too far out.
Had I a fair ledge underneath I should have felt perfectly at ease, but on the stability of the snow-wedge
I dared not calculate. To retreat was dangerous,
to advance useless ; for right in front of me was
a sheer smooth precipice, which completely extinguished the thought of further rock-work. I examined the place below me, and saw that a slip
would be attended with the very worst consequences.
To loose myself from the crag and attach myself to
the snow was so perilous an operation that I did
not attempt it ; and at length I ignobly called to
Lauener to lend me a hand. A gleam of satisfaction crossed his features as he eyed me on my
perch. He manifestly enjoyed being called to the
rescue, and exhorted me to keep quite still. He
worked up towards me, and in less than half an
hour had hold of one of my legs. 'The place is
not so bad after all,' he remarked, evidently glad
to take me down, in more senses than one. I

descended in his steps, and rejoined Hawkins upon
the snow.   From that moment Lauener was a rege-
nerate man ; the despair of his visage vanished, and
I firmly believe that the triumph he enjoyed, by
augmenting his self-respect, was the proximate cause
of our subsequent success.

The couloir was a most singular one; it was
excessively steep, and along it were two great scars,
resembling the deep-cut channels of a mountain
stream.   They were, indeed, channels cut by the
snow-torrents from the heights.   We scanned those
heights.   The view was bounded by a massive
cornice, from which the avalanches are periodically
let loose.[1]   The cornice seemed firm ; still we cast
about for some piece of rock which might shelter us
from the destroyer should he leap from his lair.
Apart from the labour of the ascent, which is great,
the frequency of avalanches will always render this
pass a dangerous one.   At 2 P.M. the air became
intensely cold.   My companion had wisely pocketed
a pair of socks, which he drew over his gloves, and
found very comforting.   My leather gloves, being
saturated with wet, were very much the reverse.

The wind was high, and as it passed the crest
of the Breithorn its moisture was precipitated and

[1] Hence the name 'Lauwinen-Thor,' which, with the consent of
Mr. Hawkins, if not at his suggestion, I have given to the pass. [The
name has since been adopted in all the maps.   March 1871.]

afterwards carried away. The clouds thus generated
shone for a time with the lustre of pearls; but as
they approached the sun they became suddenly
flooded with the most splendid iridescences.[1] At our
right now was a vertical wall of brown rock, along
the base of which we advanced. At times we were
sheltered by it, but not always; for the wind was
as fitful as a maniac, and eddying round the corners
sometimes shook us forcibly, chilled us to the mar-
row, and spit frozen dust in our faces. The snow,
moreover, adjacent to the rock had been melted and
refrozen to a steep slope of compact ice. The men
were very weary, the hewing of the steps exhausting,
and the footing, particularly on some glazed rocks
over which we had to pass, exceedingly insecure.
Once on trying to fix my alpenstock I found that
it was coated with an enamel of ice, and slipped
through my wet gloves. This startled me, for the
staff is my sole trust under such circumstances. The
crossing of those rocks was a most awkward piece
of work; a slip was imminent, and the effects of the
consequent glissade not to be calculated. We cleared
them, however, and now observed the grey haze
creeping down from the peak of the Breithorn to
the point at which we were aiming. This, however,
was visibly nearer; and, for the first time since we
began to climb, Lauener declared that he had good

[1] See 'Note on Clouds,' p. 82.

hopes—'Jetzt habe ich gute Hoffnung.' Another
hour brought us to a place where the gradient
slackened suddenly. The real work was done, and
ten minutes further wading through the deep snow
placed us fairly on the summit of the col.
Looked at from the top the pass will seem very
formidable to the best of climbers; to an ordinary
eye it would appear simply terrific. We reached
the base of the barrier at nine A.M.; we had surmounted it at four; seven hours consequently had
been spent upon that tremendous wall. Our view
was limited above; clouds were on all the mountains,
and the Great Aletsch glacier was hidden by dense
fog. With long swinging strides we went down the
slope. Several times during our descent the snow
coating was perforated, and hidden crevasses revealed.
At length we reached the glacier, and plodded along
it through the dreary fog. We cleared the ice just
at nightfall, passed the Märjelin See, and soon
found ourselves in utter darkness on the spurs of
the Æggischhorn. We lost the track and wandered
for a time bewildered. We sat down to rest, and
then learned that Lauener was extremely ill. To
quell the pangs of toothache he had chewed a cigar,
which after his day's exertion was too much for him.
He soon recovered, however, and we endeavoured to
regain the track. In vain. The guides shouted, and

after many repetitions we heard a shout in reply   A
herdsman approached, and conducted us to some
neighbouring chalets, whence he undertook the
office of guide. After a time he also found himself
in difficulty. We saw distant lights, and Lauener
once more pierced the air with his tremendous
whoop. We were heard. Lights were sent towards
us, and an additional half-hour placed us under
the roof of Herr Wellig, the active and intelligent
proprietor of the Jungfrau hotel.

After this day's journey, which was a very hard
one, the tide of health set steadily in. I have no
remembrance of any further exhibition of the symp-
toms which had driven me to Switzerland. Each
day's subsequent exercise made both brain and
muscles firmer. We remained at the Æggischhorn
for several days, occupying ourselves principally
with observations and measurements on the Aletsch
glacier, and joining together afterwards in that day's
excursion—unparalleled in my experience—which
has found in my companion a narrator worthy of its
glories. And as we stood upon the savage ledges of
the Matterhorn, with the utmost penalty which the
laws of falling bodies could inflict at hand, I felt
that there were perils at home for intellectual men
greater even than those which then surrounded us
—foes, moreover, which inspire no manhood by their
attacks, but shatter alike the architect and his house

by the same slow process of disintegration.[1] After the discipline of the Matterhorn, the fatal slope of the Col du Géant, which I visited a few days afterwards, looked less formidable than it otherwise might have done. From Courmayeur I worked round to Chamouni by Chapieu and the Col de Bonhomme. I attempted to get up Mont Blanc to visit the thermometers which I had planted on the summit a year previously; and succeeded during a brief interval of fair weather in reaching the Grands Mulets. But the gleam which tempted me thus far proved but a temporary truce to the war of elements, and, after remaining twenty hours at the Mulets, I was obliged to beat an inglorious retreat.—*Vacation Tourists*, 1860.

[1] This, I believe, was in allusion to the death of Sir Charles Barry.—J. T., 1871.

II.

DISASTER ON THE COL DU GÉANT.

On the 18th of August, while Mr. Hawkins and I were staying at Breuil, rumours reached us of a grievous disaster which had occurred on the Col du Géant. At first, however, the accounts were so contradictory as to inspire the hope that they might be grossly exaggerated or altogether false. But more definite intelligence soon arrived, and before we quitted Breuil it had been placed beyond a doubt that three Englishmen, with a guide named Tairraz, had perished on the col. On the 21st I saw the brother of Tairraz at Aosta, and learned from the saddened man all that he knew. What I then heard only strengthened my desire to visit the scene of the catastrophe, and obtain by actual contact with the facts truer information than could possibly be conveyed to me by description. On the afternoon of the 22nd I accordingly reached Courmayeur, and being informed that M. Curie, the resident French pastor, had visited the place and made an accurate sketch of it, I immediately called upon him. With the most obliging promptness he

placed his sketches in my hands, gave me a written
account of the occurrence, and volunteered to accompany me.   I gladly accepted this offer, and early
on the morning of Thursday the 21st of August we
walked up to the pavilion which it had been the
aim of the travellers to reach on the day of their
death.   Wishing to make myself acquainted with
the entire line of the fatal glissade, I walked
directly from the pavilion to the base of the rocky
couloir along which the travellers had been precipitated, and which had been described to me as so
dangerous that a chamois-hunter had declined
ascending it some days before.   At Courmayeur,
however, I secured the services of a most intrepid
man, who had once made the ascent, and who now
expressed his willingness to be my guide.   We
began our climb at the very bottom of the couloir,
while M. Curie, after making a circuit, joined us on
the spot where the body of the guide Tairraz had
been arrested, and where we found sad evidences
of his fate.   From this point onward M. Curie
shared the dangers of the ascent, until we reached
the place where the rocks ended and the fatal snow-
slope began.   Among the rocks we had frequent
and melancholy occasion to assure ourselves that we
were on the exact track.   We found there a pen-
knife, a small magnetic compass, and many other
remnants of the fall.

At the bottom of the snow-slope M. Curie quitted me, urging me not to enter upon the slope, but to take to a stony ridge on the right. No mere inspection, however, could have given me the desired information. I asked my guide whether he feared the snow, and, his reply being negative, we entered upon it together, and ascended it along the furrow which still marked the line of fall. Under the snow, at some distance up the slope, we found a fine new ice-axe, the property of one of the guides. We held on to the track up to the very summit of the col, and as I stood there and scanned the line of my ascent a feeling of augmented sadness took possession of me. There seemed no sufficient reason for this terrible catastrophe. With ordinary care the slip might in the first instance have been avoided, and with a moderate amount of skill and vigour the motion, I am persuaded, might have been arrested soon after it had begun.

Bounding the snow-slope to the left was the ridge along which travellers to Courmayeur usually descend. It is rough, but absolutely without danger. The party were, however, tired when they reached this place, and to avoid the roughness of the ridge they took to the snow. The inclination of the slope above was moderate; it steepened considerably lower down, but its steepest portion did not much exceed forty-five degrees of inclination. At all events, a skilful mountaineer might throw himself prostrate on the slope with perfect reliance on his power to arrest his downward motion.

It is alleged that when the party entered the summit of the col on the Chamouni side the guides proposed to return, but the Englishmen persisted in going forward. One thing alone could justify the proposition thus ascribed to the guides by their friends—a fog so thick as to prevent them from striking the summit of the col at the proper point, and to compel them to pursue their own traces backwards. The only part of the col hitherto regarded as dangerous had been passed, and, unless for the reason assumed, it would have been simply absurd to recross this portion instead of proceeding to Courmayeur. It is alleged that a fog existed; but the summit had been reached, and the weather cleared afterwards. Whether, therefore, the Englishmen refused to return or not on the Montanvert side, it ought in no way to influence our judgment of what occurred on the Courmayeur side, where the weather which prompted the proposal to go back ceased to be blameable.

A statement is also current to the effect that the travellers were carried down by an avalanche. In connection with this point M. Curie writes to me thus: 'Il paraît qu'à Chamounix on répand le bruit que c'est une *avalanche* qui a fait périr les

voyageurs. C'est là une fausseté que le premier vous
saurez démentir sur les lieux.' I subscribe without
hesitation to this opinion of M. Curie. That a considerable quantity of snow was brought down by the
rush was probable, but an avalanche properly so called
there was not, and it simply leads to misconception
to introduce the term at all.

We are now prepared to discuss the accident. The
travellers, it is alleged, reached the summit of the
col in a state of great exhaustion, and it is certain
that such a state would deprive them of the caution
and firmness of tread necessary in perilous places.
But a knowledge of this ought to have prevented the
guides from entering upon the snow-slope at all. We
are, moreover, informed that even on the gentler portion of the slope one of the travellers slipped repeatedly. On being thus warned of danger, why did not
the guides quit the snow and resort to the ridge?
They must have had full confidence in their power
to stop the glissade which seemed so imminent, or
else they were reckless of the lives they had in
charge. At length the fatal slip occurred, where the
fallen man, before he could be arrested, gathered
sufficient momentum to jerk the man behind him off
his feet, the other men were carried away in succession, and in a moment the whole of them were rushing downwards. What efforts were made to check
this fearful rush, at what point of the descent the

two guides relinquished the rope, which of them
gave way first, the public do not know, though this
ought to be known. All that is known to the
public is that the two men who led and followed the
party let go the rope and escaped, while the three
Englishmen and Tairraz went to destruction. Tairraz
screamed, but, like Englishmen, the others met
their doom without a word of exclamation.

At the bottom of the slope a rocky ridge, forming
the summit of a precipice, rose slightly above the
level of the snow, and over it they were tilted. I do
not think a single second's suffering could have been
endured. During the wild rush downwards the be-
wilderment was too great to permit even of fear, and
at the base of the precipice life and feeling ended
suddenly together. A steep slope of rocks connected
the base of this precipice with the brow of a second
one, at the bottom of which the first body was found.
Another slope ran from this point to the summit of
another ledge, where the second body was arrested,
while attached to it by a rope, and quite overhanging
the ledge, was the body of the third traveller. The
body of the guide Tairraz was precipitated much
further, and was much more broken.

The question has been raised whether it was right
under the circumstances to tie the men together. I
believe it was perfectly right, if only properly done.
But the actual arrangement was this: The three

Englishmen were connected by a rope tied firmly
round the waist of each of them; one end of the rope
was held in the hand of the guide who led the party;
the other end was similarly held by the hindmost
guide, while Tairraz grasped the rope near its middle.
Against this mode of attachment I would enter
an emphatic protest. It, in all probability, caused
the destruction of the unfortunate Russian traveller
on the Findelen Glacier last year, and to it I believe
is to be ascribed the disastrous issue of the slip on
the Col du Géant. Let me show cause for this protest. At a little depth below the surface the snow
upon the fatal slope was firm and consolidated, but
upon it rested a superficial layer, about ten inches
or a foot in depth, partly fresh, and partly disintegrated by the weather. By the proper action of
the feet upon such loose snow, its granules are made
to unite so as to afford a secure footing; but when
a man's body, presenting a large surface, is thrown
prostrate upon a slope covered with such snow, the
granules act like friction wheels, offering hardly any
resistance to the downward motion.

A homely illustration will render intelligible the
course of action necessary under such conditions.
Suppose a boy placed upon an oilcloth which covers
a polished table, and the table tilted to an angle of
forty-five degrees. The oilcloth would evidently
slide down, carrying the boy along with it, as the

loose snow slid over the firm snow on the Col du
Géant. But suppose the boy provided with a stick
spiked with iron, what ought he to do to check his
motion? Evidently drive his spike through the
oilcloth and anchor it firmly in the wood underneath.
A precisely similar action ought to have been resorted to on the Col du Géant. Each man as he fell
ought to have turned promptly on his face, pierced
with his armed staff the superficial layer of soft
snow, and pressed with both hands the spike into
the consolidated mass underneath. He would thus
have applied a break, sufficient not only to bring
himself to rest, but, if well done, sufficient, I believe,
to stop a second person. I do not lightly express this
opinion : it is founded on varied experience upon
slopes at least as steep as that under consideration.

Consider now the bearing of the mode of attachment above described upon the question of rescue.
When the rope is fastened round the guide's waist,
both his arms are free, to drive, in case of necessity,
his spiked staff into the snow. But in the case before us, one arm of each guide was virtually powerless ; on it was thrown the strain of the falling man
in advance, by which it was effectually fettered.
But this was not all. When the attached arm receives the jerk, the guide instinctively grasps the
rope with the other hand ; in doing so, he relin-
quishes his staff, and thus loses the sheet-anchor of

salvation. Such was the case with the two guides who escaped on the day now in question. The one lost his bâton, the other his axe, and they probably had to make an expert use of their legs to save even themselves from destruction. Tairraz was in the midst of the party. Whether it was in his power to rescue himself or not, whether he was caught in the coil of the rope or laid hold of by one of his companions, we can never know. Let us believe that he clung to them loyally, and went with them to death sooner than desert the post of duty.

III.

THE MATTERHORN—FIRST ASSAULT.

By F. VAUGHAN HAWKINS, M.A.[1]

THE summer and autumn of 1860 will long be remembered in Switzerland as the most ungenial and disastrous season, perhaps, on record; certainly without a parallel since 1834. The local papers were filled with lamentations over 'der ewige Südwind,' which overspread the skies with perpetual cloud, and from time to time brought up tremendous storms, the fiercest of which, in the three first days of September, carried away or blocked up for a time, I believe, every pass into Italy except the Bernina. At Andermatt, on the St. Gothard, we were cut off for two days from all communications whatever by water on every side. The whole of the lower Rhone valley was under water. A few weeks later, I found the Splügen, in the gorge above Chiavenna, alto-

---

[1] Instead of attempting to write one myself, I requested the permission of my friend Mr. Hawkins to republish his admirable account of our first assault upon the Matterhorn. I have to thank both him and Mr. Macmillan for the obliging promptness with which my request was granted.

gether gone, remains of the old road being just visible here and there, but no more. In the Valteline, I found the Stelvio road in most imminent danger, gangs of men being posted in the courses of the torrents to divert the boulders, which every moment threatened to overwhelm the bridges on the route. A more unlucky year for glacier expeditions, therefore, could hardly be experienced; and the following pages present in consequence only the narrative of an unfinished campaign, which it is the hope of Tyndall and myself to be able to prosecute to a successful conclusion early next August.

I had fallen in with Professor Tyndall on the Basle Railway, and a joint plan of operations had been partly sketched out between us, to combine to some extent the more especial objects of each— scientific observations on his part; on mine, the exploration of new passes and mountain topography; but the weather sadly interfered with these designs. After some glacier measurements had been accomplished at Grindelwald, a short spell of fair weather enabled us to effect a passage I had long desired to try, from Lauterbrunnen direct to the Æggischhorn by the Roththal, a small and unknown but most striking glacier valley, known to Swiss mythology as the supposed resort of condemned spirits. We scaled, by a seven hours' perpendicular climb, the vast amphitheatre of rock which bounds the

Aletsch basin on this side, and had the satisfaction
of falsifying the predictions of Ulrich Lauener, who
bade us farewell at Grindelwald with the dis-
couraging assertion that he should see us back
again, as it was quite impossible to get over where
we were going. As we descended the long reaches
of the Aletsch glacier, rain and mist again gathered
over us, giving to the scene the appearance of a
vast Polar sea, over the surface of which we were
travelling, with no horizon visible anywhere except
the distant line of level ice. Arrived at the
Æggischhorn, the weather became worse than ever;
a week elapsed before the measurement of the
Aletsch glacier could be completed; and we re-
luctantly determined to dismiss Bennen, who was
in waiting, considering the season too bad for high
ascents, and to push on with Christian Lauener to
the glaciers about Zinal. Bennen was in great
distress. He and I had the previous summer re-
connoitred the Matterhorn from various quarters,
and he had arrived at the conclusion that we could
in all probability ('ich beinahe behaupte') reach
the top. That year, being only just convalescent
from a fever, I had been unable to make the
attempt, and thus an opportunity had been lost
which may not speedily recur, for the mountain
was then (September 1859) almost free from snow.
Bennen had set his heart on our making the at-

tempt in 1860, and great was his disappointment at our proposed departure for Zinal. At the last moment, however, a change of plans occurred. Lauener was unwilling to proceed with us to Zinal : we resolved to give Bennen his chance : the theodolite was packed up and despatched to Geneva, and we set off for Breuil, to try the Matterhorn.

Accessible or not, however, the Mont Cervin is assuredly a different sort of affair from Mont Blanc or Monte Rosa, or any other of the thousand and one summits which nature has kindly opened to man, by leaving one side of them a sloping plain of snow, easy of ascent, till the brink of the precipice is reached which descends on the other side. The square massive lines of terraced crags which fence the Matterhorn, stand up on all sides nearly destitute of snow, and where the snow lies thinly on the rocks it soon melts and is hardened again into smooth glassy ice, which covers the granite slabs like a coat of varnish, and bids defiance to the axe. Every step of the way lies between two precipices, and under toppling crags, which may at any moment bring down on the climber the most formidable of Alpine dangers—a fire of falling stones. The mountain too has a sort of *prestige* of invincibility which is not without its influence on the mind, and almost leads one to expect to encounter some new

and unheard-of source of peril upon it: hence I
suppose it is that the dwellers at Zermatt and in
Val Tournanche have scarcely been willing to attempt to set foot upon the mountain, and have left
the honour of doing so to a native of another district,
who, as he has been the first mortal to plant foot on
the hitherto untrodden peak, so he will, I hope,
have the honour, which he deserves, of being the
first to reach the top.

John Joseph Bennen, of Laax, in the Upper
Rhine Valley, is a man so remarkable that I cannot
resist the desire (especially as he cannot read
English) to say a few words about his character.
Born within the limits of the German tongue, and
living amidst the mountains and glaciers of the
Oberland, he belongs by race and character to a
class of men of whom the Laueners, Melchior
Anderegg, Bortis, Christian Almer, Peter Bohren,
are also examples—a type of mountain race, having
many of the simple heroic qualities which we associate, whether justly or unjustly, with Teutonic
blood, and essentially different from—to my mind,
infinitely superior to—the French-speaking, versatile,
wily Chamouniard. The names I have mentioned are
all those of first-rate men ; but Bennen, as (I believe) he surpasses all the rest in the qualities which
fit a man for a leader in hazardous expeditions,
combining boldness and prudence with an ease and

power peculiar to himself, so he has a faculty of conceiving and planning his achievements, a way of concentrating his mind upon an idea, and working out his idea with clearness and decision, which I never observed in any man of the kind, and which makes him, in his way, a sort of Garibaldi. Tyndall, on the day of our expedition, said to him, ' Sie sind der Garibaldi der Führer, Bennen ; ' to which he answered in his simple way, ' Nicht wahr ? ' (' Am I not ? '), an amusing touch of simple vanity, a dash of pardonable bounce, being one of his not least amiable characteristics. Thoroughly sincere and ' einfach ' in thought and speech, devoted to his friends, without a trace of underhand self-seeking in his relations to his employers, there is an independence about him, a superiority to most of his own class, which makes him, I always fancy, rather an isolated man ; though no one can make more friends wherever he goes, or be more pleasant and thoroughly cheerful under all circumstances. But he left his native place, Steinen, he told me, the people there not suiting him ; and in Laax, where he now dwells, I guess him to be not perhaps altogether at home. Unmarried, he works quietly most of the year at his trade of a carpenter, unless when he is out alone, or with his friend Bortis (a man seemingly of reserved and uncommunicative disposition, but a splendid mountaineer), in the chase after chamois, of which he is passionately fond, and will tell stories, in his simple and emphatic way, with the greatest enthusiasm. Pious he is, and observant of religious duties, but without a particle of the 'mountain gloom,' respecting the prevalence of which among the dwellers in the High Alps Mr. Ruskin discourses poetically, but I am myself rather incredulous. A perfect nature's gentleman, he is to me the most delightful of companions ; and though no 'theory' defines our reciprocal obligations as guide and employer, I am sure that no precipice will ever engulf me so long as Bennen is within reach, unless he goes into it also—an event which seems impossible—and I think I can say I would, according to the measure of my capacity, do the same by him. But any one who has watched Bennen skimming along through the mazes of a crevassed glacier, or running like a chamois along the side of slippery ice-covered crags, axe and foot keeping time together, will think that—as Lauener said of his brother Johann, who perished on the Jungfrau, *he* could never fall—nothing could bring him to grief but an avalanche.[1]

[1] As Bennen and Tyndall were going up the Finsteraarhorn once upon a time, the work being severe, Bennen turned round and said to Tyndall, 'Ich fühle mich jetzt ganz wie der Tyroler einmal,' and went on to relate a story of the conversation between a priest and an honest Tyrolese, who complained to his father confessor that religion and an extreme passion for the fair sex struggled

D

Delayed in our walk from the Æggischhorn by the usual severity of the weather, Tyndall, Bennen, and myself reached Breuil on Saturday, August 18, to make our attempt on the Monday. As we approached the mountain, Bennen's countenance fell visibly, and he became somewhat gloomy; the mountain was almost white with fresh-fallen snow. 'Nur der Schnee furcht mich,' he replied to our enquiries. The change was indeed great from my recollection of the year before; the well-marked, terrace-like lines along the south face, which are so well given in Mr. George Barnard's picture, were now almost covered up; through the telescope could be seen distinctly huge icicles depending from the crags, the lines of melting snow, and the dark patches which we hoped might spread a great deal faster than they were likely to, during the space of twenty-four hours. There was nothing for it, although our prospects of success were materially diminished by the snow, but to do the best we could. As far as I was concerned, I felt that I should be perfectly satisfied with getting part of the way up on a first trial, which would make one acquainted with the nature of the rocks, dispel the

within him, and neither could expel the other. 'Mein Sohn,' said the priest, Frauen zu lieben und im Himmel zu kommen, das geht nicht.' 'Herr Pfarrer,' sagte der Tyroler, ' *es muss gehen.*' ' Und so sag' ich jetzt,' cried Bennen. ' Es muss gehen ' is always his motto.

prestige which seemed to hang over the untrodden mountain, and probably suggest ways of shortening the route on another occasion.

We wanted some one to carry the knapsack containing our provisions; and on the recommendation of the landlord at Breuil, we sent for a man, named Carrel, who, we were told, was the best mountaineer in Val Tournanche, and the nephew of M. le Chanoine Carrel, whose acquaintance I once had the honour of making at Aosta. From the latter description I rather expected a young, and perhaps aristocratic-looking personage, and was amused at the entrance of a rough, good-humoured, shaggy-breasted man, between forty and fifty, an ordinary specimen of the peasant class. However, he did his work well, and with great good temper, and seemed ready to go on as long as we chose; though he told me he expected we should end by passing the night somewhere on the mountain, and I don't think his ideas of our success were ever very sanguine.

We were to start before 3 A.M. on Monday morning, August 20; and the short period for sleep thus left us was somewhat abridged in my own case, not so much by thoughts of the coming expedition, as by the news which had just reached us in a vague, but, unfortunately, only too credible form, of the terrible accident on the Col du Géant a few days before. The account thus reaching us was naturally

magnified, and we were as yet ignorant of the names.
I could not at night shake off the (totally groundless)
idea that a certain dear friend of mine was among
them, and that I ought at that moment to be hurry-
ing off to Courmayeur, to mourn and to bury him.
In the morning, however, these things are forgotten;
we are off, and Carrel pilots us with a lantern across
the little stream which runs by Breuil, and up the
hills to the left, where in the darkness we seem from
the sound to be in the midst of innumerable rills of
water, the effects of the late rains.   The dark outline
of the Matterhorn is just visible against the sky, and
measuring with the eye the distance subtended by
the height we have to climb, it seems as if success
must be possible : so hard is it to imagine all the
ups and downs which lie in that short sky-line.

Day soon dawns, and the morning rose-light
touches the first peak westward of us; the air is
wonderfully calm and still, and for to-day, at all
events, we have good weather, without that bitter
enemy the north wind ; but a certain opaque look in
the sky, long streaks of cloud radiating from the
south-west horizon up towards the zenith, and the
too dark purple of the hills south of Aosta, are signs
that the good weather will not be lasting.   By five
we are crossing the first snow-beds, and now Carrel
falls back, and the leader of the day comes to the
front : all the day he will be cutting steps, but those

compact and powerful limbs of his will show no signs
of extra exertion, and to-day he is in particularly
good spirits. Carpentering, by the way—not fine
turning and planing, but rough out-of-doors work,
like Bennen's—must be no bad practice to keep
hand and eye in training during the dead season.
We ascend a narrow edge of snow, a cliff some way
to the right: the snow is frozen and hard as rock,
and arms and legs are worked vigorously. Tyndall
calls out to me, to know whether I recollect the
' conditions:' *i.e.* if your feet slip from the steps,
turn in a moment on your face, and dig in hard
with alpenstock in both hands under your body; by
this means you will stop yourself if it is possible.
Once on your back, it is all over, unless others
can save you: you have lost all chance of helping
yourself. In a few minutes we stop, and rope all
together, in which state we continued the whole
day. The prudence of this some may possibly
doubt, as there were certainly places where the
chances were greater that if one fell, he would drag
down the rest, than that they could assist him; but
we were only four, all tolerably sure-footed, and in
point of fact I do not recollect a slip or stumble of
consequence made by any one of us. Soon the slope
lessens for a while, but in front a wall of snow
stretches steeply upwards to a gap, which we have
to reach, in a kind of recess, flanked by crags of

formidable appearance. We turn to the rocks on
the left hand. As, to one walking along miry
ways, the opposite side of the path seems ever the
most inviting, and he continually shifting his course
from side to side lengthens his journey with small
profit, so in ascending a mountain one is always
tempted to diverge from snow to rocks, or *vice
versâ*. Bennen had intended to mount straight up
towards the gap, and it is best not to interfere with
him; he yields, however, to our suggestions, and we
assail the rocks. These, however, are ice-bound,
steep, and slippery: hands and knees are at work,
and progress is slow. At length we stop upon a
ledge where all can stand together, and Carrel
proposes to us (for Bennen and he can only commu-
nicate by signs, the one knowing only French, the
other German) to go on and see whether an easier
way can be found still further to the left. Bennen
gives an approving nod: he looks with indulgent
pity on Carrel, but snubs all remarks of his as to
the route. ' Er weiss gar nichts,' he says. Carrel
takes his axe, and mounts warily, but with good
courage; presently he returns, shaking his head.
The event is fortunate, for had we gone further to
the left we should have reached the top of the
ridge from which, as we afterwards found, there is
no passage to the gap, and our day's work would
probably have ended then and there. Bennen now

leads to the right, and moves swiftly up from ledge
to ledge. Time is getting on, but at length we
emerge over the rocks just in face of the gap,
and separated from it by a sort of large snow-crater,
overhung on the left by the end of the ridge, from
which stones fall which have scarred the sides of the
crater. The sides are steep, but we curve quickly
and silently round them : no stones fall upon us ;
and now we have reached the narrow neck of snow
which forms the actual gap; it is half-past eight,
and the first part of our work is done.

By no means the hardest part, however. We
stand upon a broad red granite slab, the lowest step
of the actual peak of the Matterhorn : no one has
stood there before us. The slab forms one end of
the edge of snow, surmounted at the other end by
some fifty feet of overhanging rock, the end of the
ridge. On one side of us is the snow-crater, round
which we had been winding; on the other side a
scarped and seamed face of snow drops sheer on the
north, to what we know is the Zmutt glacier. Some
hopes I had entertained of making a pass by this
gap, from Breuil to Zermatt, vanish immediately.
Above us rise the towers and pinnacles of the Matterhorn, certainly a tremendous array. Actual contact
immensely increases one's impressions of this, the
hardest and strongest of all the mountain masses of
the Alps ; its form is more remarkable than that of

other mountains, not by chance, but because it is
built of more massive and durable materials, and
more solidly put together: nowhere have I seen
such astonishing masonry. The broad gneiss blocks
are generally smooth and compact, with little appearance of splintering or weathering. Tons of
rock, in the shape of boulders, must fall almost daily
down its sides, but the amount of these, even in the
course of centuries, is as nothing compared with the
mass of the mountain; the ordinary processes of
disintegration can have little or no effect on it. If
one were to follow Mr. Ruskin, in speculating on
the manner in which the Alpine peaks can have
assumed their present shape, it seems as if such a
mass as this can have been blocked out only while
rising from the sea, under the action of waves such
as beat against the granite headlands of the Land's
End. Once on dry land, it must stand as it does
now, apparently for ever.

Two lines of ascent offer, between which we have
to choose: one along the middle or dividing ridge,
the back-bone of the mountain, at the end of which
we stand; the other by an edge some little way to
the right: a couloir lies between them. We choose
the former, or back-bone ridge; but the other
proves to be less serrated, and we shall probably try
it on another occasion. As we step from our halting-place, Bennen turns round and addresses us in

a few words of exhortation, like the generals in
Thucydides. He knows us well enough to be sure
that we shall not feel afraid, but every footstep
must be planted with the utmost precaution : no
fear, 'wohl immer Achtung.' Soon our difficulties
begin; but I despair of relating the incidents of
this part of our route, so numerous and bewildering
were the obstacles along it; and the details of each
have somewhat faded from the memory. We are
immersed in a wilderness of blocks, roofed and
festooned with huge plates and stalactites of ice, so
large that one is half disposed to seize hold and
clamber up them. Round, over, and under them
we go : often progress seems impossible ; but
Bennen, ever in advance, and perched like a bird on
some projecting crag, contrives to find a way. Now
we crawl singly along a narrow ledge of rock, with a
wall on one side, and nothing on the other : there is
no hold for hands or alpenstock, and the ledge slopes
a little, so that if the nails in our boots hold not,
down we shall go : in the middle of it a piece of
rock juts out, which we ingeniously duck under,
and emerge just under a shower of water, which
there is no room to escape from. Presently comes
a more extraordinary place : a perfect chimney of
rock, cased all over with hard black ice, about an
inch thick. The bottom leads out into space, and
the top is somewhere in the upper regions : there is

absolutely nothing to grasp at, and to this day I cannot understand how a human being could get up or down it unassisted. Bennen, however, rolls up it somehow, like a cat; he is at the top, and beckons Tyndall to advance; my turn comes next; I endeavour to mount by squeezing myself against the sides, but near the top friction suddenly gives way, and down comes my weight upon the rope:— a stout haul from above, and now one knee is upon the edge, and I am safe: Carrel is pulled up after me. After a time, we get off the rocks, and mount a slope of ice, which curves rapidly over for about three yards to our left, and then (apparently) drops at once to the Zmutt glacier. We reach the top of this, and proceed along it, till at last a sort of pinnacle is reached, from which we can survey the line of towers and crags before us up to a point just below the actual top, and we halt to rest a while. Bennen goes on to see whether it be possible to cross over to the other ridge, which seems an easier one. Left to himself, he treads lightly and almost carelessly along. ' Geb' Acht, Bennen ! ' (Take care of yourself) we shout after him, but needlessly; he stops and moves alternately, peering wistfully about, exactly like a chamois; but soon he returns, and says there is no passage, and we must keep to the ridge we are on.

Three hours had not yet elapsed since we left the

gap, and from our present station we could survey
the route as far as a point which concealed from
us the actual summit, and could see that the
difficulties before us were not greater than we had
already passed through, and such as time and perseverance would surely conquer. Nevertheless, there
is a tide in the affairs of such expeditions, and the
impression had been for some time gaining ground
with me that the tide on the present occasion had
turned against us, and that the time we could
prudently allow was not sufficient for us to reach
the top that day. Before trial, I had thought it
not improbable that the ascent might turn out
either impossible or comparatively easy; it was
now tolerably clear that it was neither the one nor
the other, but an exceedingly long and hard piece
of work, which the unparalleled amount of ice made
longer and harder than usual. I asked Bennen if
he thought there was time enough to reach the top
of all: he was evidently unwilling, however, to give
up hopes; and Tyndall said he would give no
opinion either way; so we again moved on.

At length we came to the base of a mighty knob,
huger and uglier than its fellows, to which a little
arête of snow served as a sort of drawbridge. I
began to fear lest in the ardour of pursuit we
might be carried on too long, and Bennen might
forget the paramount object of securing our safe

retreat. I called out to him that I thought I should stop somewhere here, that if he could go faster alone he might do so, but he must turn in good time. Bennen, however, was already climbing with desperate energy up the sides of the kerb; Tyndall would not be behind him; so I loosed the rope and let them go on. Carrel moved back across the little *arête*, and sat down, and began to smoke: I remained for a while standing with my back against the knob, and gazed by myself upon the scene around.

As my blood cooled, and the sounds of human footsteps and voices grew fainter, I began to realise the height and the wonderful isolation of our position. The air was preternaturally still; an occasional gust came eddying round the corner of the mountain, but all else seemed strangely rigid and motionless, and out of keeping with the beating heart and moving limbs, the life and activity of man. Those stones and ice have no mercy in them, no sympathy with human adventure; they submit passively to what man can do; but let him go a step too far, let heart or hand fail, mist gather or sun go down, and they will exact the penalty to the utter-most. The feeling of 'the sublime' in such cases depends very much, I think, on a certain balance between the forces of nature and man's ability to cope with them: if they are too weak, the scene

fails to impress; if they are too strong for him,
what was sublime becomes only terrible. Looking
at the Dôme du Goûté or the Zumstein Spitze full
in the evening sun, when they glow with an absolutely unearthly loveliness, like a city in the heavens,
I have sometimes thought that, place but the
spectator alone just now upon those shining
heights, with escape before night all but impossible,
and he will see no glory in the scene—only the
angry eye of the setting sun fixed on dark rocks and
dead-white snow.

We had risen seemingly to an immense height
above the gap, and the ridge which stretches from the
Matterhorn to the Dent d'Érin lay flat below; but
the peak still towered behind me, and, measuring our
position by the eye along the side of our neighbour
of equal height, the Weisshorn, I saw that we must
be yet a long way beneath the top. The gap itself,
and all traces of the way by which we had ascended,
were invisible; I could see only the stone where
Carrel sat, and the tops of one or two crags rising
from below. The view was, of course, magnificent,
and on three sides wholly unimpeded : with one hand
I could drop a stone which would descend to Zmutt,
with the other to Breuil. In front lay, as in a map,
the as yet unexplored peaks to south and west of the
Dent d'Érin, the range which separates Val Tournanche from the Valpelline, and the glacier region

beyond, called in Ziegler's map Zardezan, over which a pass, perchance, exists to Zermatt. An illimitable range of blue hills spread far away into Italy.

I walked along the little *arête*, and sat down; it was only broad enough for the foot, and in perfectly cold blood even this perhaps might have appeared uncomfortable. Turning to look at Tyndall and Bennen, I could not help laughing at the picture of our progress under difficulties. They seemed to have advanced only a few yards. 'Have you got no further than that yet?' I called out, for we were all the time within hearing. Their efforts appeared prodigious: scrambling and sprawling among the huge blocks, one fancied they must be moving along some unseen bale of heavy goods instead of only the weight of their own bodies. As I looked, an ominous visitant appeared: down came a fragment of rock, the size of a man's body, and dashed past me on the couloir, sending the snow flying. For a moment I thought they might have dislodged it; but looking again I saw it had passed over their heads, and come from the crags above. Neither of them, I believe, observed the monster; but Tyndall told me afterwards that a stone, possibly a splinter from it, had hit him in the neck, and nearly choked him. I looked anxiously again, but no more followed. A single shot, as it were, had been fired across our bows; but the ship's course was already on the point of being put about.

Expecting fully that they would not persevere beyond a few minutes longer, I called out to Tyndall to know how soon they meant to be back. 'In an hour and a half,' he replied, whether in jest or earnest, and they disappeared round a projecting corner. A sudden qualm seized me, and for a few minutes I felt extremely uncomfortable : what if the ascent should suddenly become easier, and they should go on, and reach the top without me? I thought of summoning Carrel, and pursuing them; but the worthy man sat quietly, and seemed to have had enough of it. My suspense, however, was not long : after two or three minutes the clatter, which had never entirely ceased, became louder, and their forms again appeared : they were evidently descending. In fact, Bennen had at length turned, and said to Tyndall, 'Ich denke die Zeit ist zu kurz.' I was glad that he had gone on as long as he chose, and not been turned back on my responsibility. They had found one part of this last ascent the worst of any, but the way was open thenceforward to the farthest visible point, which can be no long way below the actual top.

It was now just about mid-day, and ample time for the descent, in all probablity, was before us ; but we resolved not to halt for any length of time till we should reach the gap. Descending, unlike ascending, is generally not so bad as it seems ; but in some places here only one can advance at a time, the

other carefully holding the rope. 'Tenez forte-
ment, Carrel, tenez,' is constantly impressed on the
man who brings up the rear. 'Splendid practice for
us, this,' exclaims Tyndall exultingly, as each successive difficulty is overcome. At length we reach a
place whence no egress is possible; we look in vain
for traces of the way we had come: it is our friend
the ice-coated chimney. Bennen gets down first, in
the same mysterious fashion as he got up, and assists
us down; presently a shout is heard behind; Carrel
is attempting to get down by himself, and has stuck
fast; Bennen has to extricate him. We are now
getting rapidly lower; soon the difficulties diminish;
our gap appears in sight, and once more we reach
the broad granite slab beside the narrow col, and
breathe more freely.

Two hours have brought us down thus far; but if
we are to return by the way we came, three or four
hours of hard work are still needed before we arrive
at anything like ordinary snow-walking. We hold
a consultation. Bennen thinks the rocks, now that
the ice is melting in the afternoon sun, will be
difficult, and 'withal somewhat dangerous' (etwa
gefährlich auch). The reader will remark that
Bennen uses the word 'dangerous' in its legitimate
sense. A place is dangerous where a good climber
cannot be secure of his footing; a place is *not*
dangerous where a good climber is in no danger of

slipping, although to slip may be fatal. We determine to see if it be possible to descend the sides of
the snow-crater, on the brink of which we now stand.
The crater is portentously steep, deeply lined with
fresh snow, which glistens and melts in the powerful
sun. The experiment is slightly hazardous, but we
resolve to try. The crater appears to narrow gradually to a sort of funnel far down below, through
which we expect to issue into the glacier beneath.
At the sides of the funnel are rocks, which some one
suggests might serve to break our fall, should the
snow go down with us, but their tender mercies seem
to me doubtful. Cautiously, with steady, balanced
tread, we commit ourselves to the slope, distributing
the weight of the body over as large a space of snow
as possible, by fixing in the pole high up, and the
feet far apart, for a slip or stumble now will probably dissolve the adhesion of the fresh, not yet
compacted mass, and we shall go down to the bottom
in an avalanche. Six paces to the right, then again
to the left; we are at the mercy of those overhanging rocks just now, and the recent tracks of stones
look rather suspicious; but all is silent, and soon
we gain confidence, and congratulate ourselves on
an expedient which has saved us hours of time and
toil. Just to our right the snow is sliding by, first
slowly, then faster; keep well out of the track of
it, for underneath is a hard polished surface, and

E

if your foot chance to light there, off you will probably shoot. The snow travels much faster than we
do, or have any desire to do; we are like a coach
travelling alongside of an express train; in popular
phrase, we are going side by side with a small avalanche, though a real avalanche is a very different
matter. Soon we come somewhat under the lee of the
rocks, and now all risk is over, we are through the
funnel, and floundering waist-deep, heedless of crevasses in the comparatively level slopes beyond. We
plunge securely down now in the deep snow, where
care and caution had been requisite in crossing the
frozen surface in the morning; at length we cast
off the rope, and are on terra firma.

We shall be at Breuil in unexpectedly good time,
before five o'clock; but it is well we are off the
mountain early, for clouds and mist are already
gathering round the peak, and the weather is about
to break. Tyndall rushes rapidly down the slopes,
and is lost to view; Bennen and I walk slowly, discussing the results of the day. I am glad to see
that he is in high spirits, and confident of our future
success. He agrees with me to reach the top will
be an exceedingly long day's work, and that we must
allow ten hours at least for the actual peak, six to
ascend, four to descend; we must start next time,
he says, 'ganz, ganz früh,' and manage to reach the
gap by seven o'clock. Presently we deviate a little

from our downward course ; the same thought occu-
pies our minds ; we perceive a long low line of roof on
the mountain-side, and are not mistaken in supposing
that our favourite food will at this hour be found there
in abundance. The shepherds on the Italian hills
are more hospitable and courteous, I think, than their
Swiss brethren : twenty cows are moving their tails
contentedly in line under the shed, for Breuil is a
rich pasture valley, and in an autumn evening I
have counted six herds of from ninety to a hundred
each, in separate clusters, like ants, along the stream
in the distance. The friendly man, in hoarse but
hearty tones, urges us on as we drink ; Bennen puts
into his hand forty centimes for us both (for we
have disposed of no small quantity): but he is with
difficulty persuaded to accept so large a sum, and
calls after us, ' C'est trop, c'est trop, messieurs.'
Long may civilisation and half-francs fail of reaching his simple abode ; for, alas ! the great tourist-
world is corrupting the primitive chalet-life of the
Alps, and the Alpine man returning to his old
haunts, finds a rise in the price even of ' niedl ' and
' mascarpa.'

The day after our expedition Bennen and myself
recrossed the Théodule in a heavy snow-storm.
Tyndall started for Chamouni, for the weather was
too bad to justify an indefinite delay at Breuil in
the hope of making another attempt that year, and

by waiting till another season we were sure of obtaining less unfavourable conditions of snow and ice upon the mountain.—We had enjoyed an exciting and adventurous day, and I myself was not sorry to have something still left to do, while we had the satisfaction of being the first to set foot on this, the most imposing and mightiest giant of the Alps—the ' inaccessible' Mont Cervin.—*Vacation Tourists*, 1860.

IV.

THERMOMETRIC STATION ON MONT BLANC.

THE thermometers referred to at p. 17 were placed on Mont Blanc in 1859. I had proposed to the Royal Society some time previously to establish a series of stations between the top and bottom of the mountain, and the council of the society was kind enough to give me its countenance and aid in the undertaking. At Chamouni I had a number of wooden piles shod with iron. The one intended for the summit was twelve feet long and three inches square; the others, each ten feet long, were intended for five stations between the top of the mountain and the bottom of the Glacier de Bossons. Each post was furnished with a small cross-piece, to which a horizontal minimum thermometer might be attached. Six-and-twenty porters were found necessary to carry all the apparatus to the Grands Mulets, whence fourteen of them were immediately sent back. The other twelve, with one exception, reached the summit, whence six of them were sent

back. Six therefore remained. In addition to
these we had three guides, Auguste Balmat being
the principal one ; these, with Dr. Frankland and
myself, made up eleven persons in all. Though the
main object of the expedition was to plant the posts
and fix the thermometers, I was very anxious to
make some observations on the transparency of the
lower strata of the atmosphere to the solar heat-
rays. I therefore arranged a series of observations
with the Abbé Veuillet, of Chamouni ; he was to
operate in the valley, while I observed at the
top. Our instruments were of the same kind ; in
this way I hoped to determine the influence of the
stratum of air interposed between the top and bottom
of the mountain upon the solar radiation.

Wishing to commence the observations at daybreak, I had a tent carried to the summit, where I
proposed to spend the night. The tent was ten feet
in diameter, and into it the whole eleven of us were
packed. The north wind blew rather fiercely over
the summit, but we dropped down a few yards to
leeward, and thus found shelter. Throughout the
night we did not suffer at all from cold, though we
had no fire, and the adjacent snow was 15° Cent.,
or 27° Fahr., below the freezing point of water.
We were all however indisposed. I was indeed very
unwell when I quitted Chamouni ; but had I fal-
tered my party would have melted away. I had

frequently cast off illness on previous occasions, and hoped to do so now. But in this I was unsuccessful; my illness was more deep-rooted than ordinary, and it augmented during the entire period of the ascent. Towards morning, however, I became stronger, while with some of my companions the reverse was the case. At daybreak the wind increased in force, and as the fine snow was perfectly dry, it was driven over us in clouds. Had no other obstacle existed, this alone would have been sufficient to render the observations on solar radiation impossible. We were therefore obliged to limit ourselves to the principal object of the expedition— the erection of the post for the thermometers. It was sunk six feet in the snow, while the remaining six feet were exposed to the air. A minimum thermometer was screwed firmly on to the crosspiece of the post; a maximum thermometer was screwed on beneath this, and under this again a wet and dry bulb thermometer. Two minimum thermometers were also placed in the snow—one at a depth of six, and the other at a depth of four feet below the surface—these being intended to give some information as to the depth to which the winter cold penetrates. At each of the other stations we placed a minimum thermometer in the ice or snow, and a maximum and a minimum in the air.

The stations were as follows:—The summit, the Corridor, the Grand Plateau, the glacier near the Grands Mulets, and two additional ones between the Grands Mulets and the end of the Glacier de Bossons. We took up some rockets, to see whether the ascensional power, or the combustion, was affected by the rarity of the air. During the night, however, we were enveloped in a dense mist, which defeated our purpose. One rocket was sent up which (though we did not know it) penetrated the mist, and was seen at Chamouni. Lecomte's experiments on the alleged influence of light and rarefaction in retarding combustion caused me to resolve on making a series of experiments on Mont Blanc. Dr. Frankland was kind enough to undertake their execution. Six candles were chosen at Chamouni, and carefully weighed. All of them were permitted to burn for one hour at the top, and were again weighed when we returned to Chamouni. They were afterwards permitted to burn an hour below. Rejecting one candle, which gave a somewhat anomalous result, we found that the quantity consumed above was, within the limits of error, the same as that consumed at the bottom. This result surprised us all the more, inasmuch as the *light* of the candles appeared to be much feebler at the top than at the bottom of the mountain.

The explosion of a pistol was sensibly weaker at

the top than at a low level. The *shortness* of the
sound was remarkable; but it bore no resemblance
to the sound of a cracker, to which in acoustic
treatises it is usually compared. It resembled more
the sound produced by the expulsion of a cork from
a champagne-bottle, but it was much louder. The
sunrise from the summit was singularly magnificent.
The snow on the shaded flanks of the mountain was
of a pure blue, being illuminated solely by the
reflected light of the sky; the summit of the mountain, on the contrary, was crimson, being illuminated
by *transmitted* light. The contrast of both was finer
than I can describe.

About twenty hours were spent upon the top of
Mont Blanc on this occasion. Had I been better
satisfied with the conduct of the guides, it would
have given me pleasure at the time to dwell upon this
out-of-the-way episode in mountain life. But a temper, new to me, and which I thought looked very like
mutiny, showed itself on the part of some of my men.
Its manifestation was slight, I must say, in most
cases, and conspicuous only in one. Regrets and
apologies followed, and due allowance ought to be
made for the perfectly novel position in which the
men found themselves. The awe of entire strange-
ness is very powerful in some minds; and to my
companions the notion of spending a night at the
top of Mont Blanc was passing strange. The thing

had never been attempted previously, nor has the experiment been repeated since.

As stated at p. 17, I made an attempt during the execrable weather of 1860 to reach the top, but was driven down after a delay of twenty hours at the Grands Mulets. The same weather destroyed the lower stations. In 1861, though the cross still remained at the top, the thermometers exhibited broken columns and were worthless for observation.

I may add, in conclusion, that the lowest temperature at the summit of the Jardin during the winter of 1858 was 21° Cent. below zero. In 1859 I vainly endeavoured to find a thermometer which had been placed in the snow upon the summit of Mont Blanc a year previously.

V.

FROM a little book called 'Mountaineering in 1861,' published nine years ago, but long since out of print, I will now make a few selections. The mountain work of that year embraced the ascent of the Weisshorn, and the passage of the barrier between the Cima di Jazzi and Monte Rosa by an untried and dangerous route. Both these expeditions are described. But, besides these narratives of outward action, I notice in the book a subjective element, consisting of the musings and reflections to which I often abandon myself when sauntering over easy ground, and without which even Switzerland would sometimes be monotonous to me. It is only from the reader accustomed to similar reflective moods that I expect acceptance, or even tolerance, of these musings: the man of action will pass them impatiently by. I begin with

A LETTER FROM BÂLE.

'I reached Bâle last night, and now sit on the balcony of the "Three Kings" with the Rhine flashing below me. It is silent here, but higher up, in passing the props of a bridge, it breaks into foam; its compressed air-bubbles burst like elastic springs, and shake the air into sonorous vibrations.[1] Thus the rude mechanical motion of the river is converted into music. The hammer of the boat-builder rings on his plank, the leaves of the poplars rustle in the breeze, the watchdog's honest bark is heard in the distance; while from the windows of the houses along the banks gleam a series of reflected suns, each surrounded by a coloured glory.

'Yesterday I travelled from Paris, and the day previous from London, when the trail of a spent storm swept across the sea and kept its anger awake. The stern of our boat went up and down, the distant craft were equally pendulous, and the usual results followed. Men's faces waxed green, roses faded from ladies' cheeks; while puzzled children yelled intermittently in the grasp of the demon which had newly taken possession of them. One rare pale maiden sat right in the line of the spray, and bore the violence of the ocean with the resignation of an angel. A white arm could be seen shining through translucent muslin, but even against it the brine beat as if it were a mere seaweed. I sat at rest, hovering fearfully on the verge of that doleful region, whose bourne most of those on board had already passed, thinking how directly materialistic

---

[1] See note at the end of this chapter.

is the tendency of sea-sickness, through its remorseless demonstration of the helplessness of the human
soul and will.

'The morning of the 1st of August found me on
my way from Paris to Bâle. The sun was strong,
and, in addition to this source of temperature, eight
human beings, each burning the slow fire which we
call life, were cooped within the limits of our compart-
ment. We slept, first singly, then by groups, and
finally as a whole. Vainly we endeavoured to ward
off the coming lethargy. Thought gradually slips
away from its object, or the object glides out of the
nerveless grasp of thought, and we are conquered by
the heat. But what *is* heat, that it should work such
changes in moral and intellectual nature? Why
are we unable to read "Mill's Logic" or study the
"Kritik der reinen Vernunft" with any profit in a
Turkish bath? Heat, defined without reference to
our sensations, is a kind of motion, as strictly
mechanical as the waves of the sea, or as the aërial
vibrations which produce sound. The communica-
tion of this motion to the molecules of the brain
produces the moral and intellectual effects just referred to. Human action is only possible within a
narrow zone of temperature. Transgress the limit
on one side, and we are torpid by excess; transgress
it on the other, and we are torpid by defect. The
intellect is in some sense a function of temperature.

Thus at noon we were drained of intellectual
energy; eight hours later the mind was awake and
active, and through her operations was shed that
feeling of earnestness and awe which the mystery of
the starry heavens ever inspires. Physically considered, however, the intellect of noon differed from
that of 8 P.M. simply in the amount of motion possessed by the molecules of the brain.

' It is not levity which prompts me to write thus.
Matter, in relation to vital phenomena, has yet to be
studied, and the command of Canute to the waves
would be wisdom itself compared with any attempt
to stop such enquiries. Let the tide rise, and let
knowledge advance; the limits of the one are not
more rigidly fixed than those of the other; and no
worse infidelity could seize upon the mind than the
belief that a man's earnest search after truth should
culminate in his perdition.'

The sun was high in heaven as we rolled away
from Bâle on the morning of the 2nd. Sooner or
later every intellectual canker disappears before
earnest work, the influence of which, moreover,
fills a wide margin beyond the time of its actual
performance. Thus, to-day, I sang as I rolled
along—not with boisterous glee, but with serene
and deep-lying gladness of heart. This happiness,
however, had its roots in the past, and had I not
been a worker previous to my release from London,

I could not now have been so glad an idler. In any other country than Switzerland the valley through which we sped would have called forth admiration and delight. Noble fells, proudly grouped, flanked us right and left. Cloud-like woods of pines overspread them in broad patches, with between them spaces of the tenderest green, while among the meadows at their feet gleamed the rushing Rhine. The zenith was blue, but the thick stratum of horizontal air invested the snowy peaks with a veil of translucent haze, through which their vast and spectral outlines were clearly seen. As we rolled on towards Thun the haze thickened, while dense and rounded clouds burst upwards, as if let loose from a prison behind the mountains. Soon afterwards the black haze and blacker clouds resolved themselves into a thunderstorm. The air was cut repeatedly by zigzag bars of solid light. Then came the cannonade, and then the heavy rain-pellets rattling with fury against the carriages. It afterwards cleared, but not wholly. Stormy cumuli swept round the mountains, between which, however, the illuminated ridges seemed to swim in the opalescent air.

At Thun I found my faithful and favourite guide, Johann Bennen, of Laax, in the valley of the Rhone, the strongest limb and stoutest heart of my acquaintance in the Alps. We took the steamer to Interlaken, and while we were on the lake the heavens again darkened, and the deck was flooded by the gushing rain. The dusky cloud-curtain was rent at intervals, and through the apertures thus formed parallel bars of extraordinary radiance escaped across the lake. On reaching Interlaken I drove to the steamer on the lake of Brientz. We started at 6 P.M., with a purified atmosphere, and passed through scenes of serene beauty in the tranquil evening light. The bridge of Brientz had been carried away by the floods, the mail was intercepted, and I joined a young Oxford man in a vehicle to Meyringen. The west wind again filled the atmosphere with gloom, and after supper I spent an hour watching the lightning thrilling behind the clouds. The darkness was intense, and the intermittent glare corespondingly impressive. Sometimes the lightning seemed to burst, like a fireball, midway between the horizon and the zenith, spreading a vast glory behind the clouds and revealing all their outlines. In front of me was a craggy summit, which indulged in intermittent shots of thunder; sharp, dry, and sudden, with scarcely an echo to soften them off.

NOTE ON THE SOUND OF AGITATED WATER.

A LIQUID vein descending through a round hole in the bottom of a tin vessel exhibits two distinct portions, the one steady and limpid, the other unsteady and apparently turbid. The flash of an electric spark in a dark room instantly resolves the turbid portion into isolated drops. Experiments made in 1849 with such a jet directed my attention to the origin of the sound of agitated water. When the smoke is projected from the lips of a tobacco-smoker, a little explosion usually occurs, which is chiefly due to the sudden bursting of the film of saliva connecting both lips. An inflated bladder bursts with an explosion as loud as a pistol-shot. Sound to some extent always accompanies the sudden liberation of compressed air, and this fact is also exhibited in the deportment of a water-jet. If the surface of water into which the jet falls intersect its limpid portion. the jet enters *silently*, and no bubbles are produced. If the surface cut the turbid portion of the jet, bubbles make their appearance with an accompaniment of sound. The very nature of the sound pronounces its origin to be the bursting of the bubbles; and to the same cause the murmur of streams and the sound of breakers appear to be almost exclusively due. The impact of water against water is a comparatively subordinate cause of the sound, and could never of itself occasion the 'babble' of a brook or the musical roar of the ocean.—*Philosophical Magazine*, February 1857.

F

VI.

THE URBACHTHAL AND GAULI GLACIER.

OUR bivouac at Meyringen was *le Sauvage*, who discharged his duty as a host with credit to himself and with satisfaction to us. Forster (the statesman) arrived, and in the afternoon of the 3rd we walked up the valley, with the view of spending the night at Hof. Between Meyringen and Hof, the vale of Hasli is crossed by a transverse ridge called the Kirchet, and the barrier is at one place split through, forming a deep chasm with vertical sides through which plunges the river Aar. The chasm is called the Finsteraarschlucht, and by the ready hypothesis of an earthquake its formation has been explained. Man longs for causes, and the weaker minds, unable to restrain their longing, often barter, for the most sorry theoretic pottage, the truth which patient enquiry would make their own. This proneness of the human mind to jump to conclusions, and thus shirk the labour of real investigation, is a most mischievous tendency. We complain of the contempt with which practical men regard theory, and,

to confound them, triumphantly exhibit the specu-
lative achievements of master minds.   But the
practical man, though puzzled, remains uncon-
vinced; and why?   Simply because nine out of ten
of the theories with which he is acquainted are de-
serving of nothing better than contempt.   Our master
minds built their theoretic edifices upon the rock of
fact, the quantity of fact necessary to enable them
to divine the *law* being a measure of individual
genius, and not a test of philosophic system.[1]

The level plain of Hof lies above the mound of
the Kirchet; how was this flat formed?   Is it not
composed of the sediment of a lake?   Did not the
Kirchet form the dam of this lake, a stream issuing
from the latter and falling over the dam?   And as
the sea-waves find a weak point in the cliffs against
which they dash, and gradually eat their way so as
to form caverns with high vertical sides, as at the
Land's End, a joint or fault or some other accidental
weakness determining their line of action; so also a
mountain torrent rushing for ages over the same dam
would be sure to cut itself a channel.   The lake
after its drainage left the basis of green meadows
as sediment behind; and through these meadows
now flows the stream of the Aar.   Imagination is

---

[1] This was written soon after Mr. Buckle's Royal Institution
lecture, which I thought a piece of astonishing rhetoric, but of very
unsound science.

essential to the natural philosopher, but its matter must be facts; and its function the discernment of their connection.

We were called at 4 A.M., an hour later than we intended, and the sight of the cloudless mountains was an inspiration to us all. At 5.30 A.M. we were off, crossing the valley of Hof, which was hugged round its margin by a light and silky mist. We ascended a spur which separated us from the Urbachthal, through which our route lay. The Aar for a time babbled in the distance, until, on turning a corner, its voice was suddenly quenched by the louder music of the Urbach, rendered mellow and voluminous by the resonance of the chasm into which the torrent leaped. The sun was already strong. His yellow light glimmered from the fresh green leaves; it smote with glory the boles and the plumes of the pines; soft shadows fell from shrub and rock upon the pastures; snow-peaks were in sight, cliffy summits also, without snow or verdure, but in many cases buttressed by slopes of soil which bore a shaggy growth of trees. To the right of us rose the bare cliffs of the Engelhörner, broken at the top into claw-shaped masses which were turned, as if in spite, against the serene heaven. Bennen walked on in front, a mass of organised force, silent, but emitting at times a whistle which sounded like the piping of a lost chamois. In a hollow of the

Engelhörner a mass of snow had found a lodgment; melted by the warm rock, its foundation was sapped, and down it came in a thundering cascade. The thick pinewoods to our right were furrowed by the tracks of these destroyers, the very wind of which, it is affirmed, tears up distant trees by the roots.

For a time our route lay through a spacious valley, which at length turned to the left, and narrowed to a gorge. Along its bottom the hissing river rushed; this we crossed, climbed the wall of a *cul de sac*, and from its rim enjoyed a glorious view. The Urbachthal has been the scene of vast glacier action. Looking at these charactered cliffs, one's thoughts involuntarily revert to the ancient days, and we restore in idea a state of things which had disappeared from the world before the development of man. Whence this wondrous power of reconstruction? Was it locked like latent heat in ancient inorganic nature, and developed as the ages rolled? Are other and grander powers still latent in nature, destined to blossom in another age? Let us question fearlessly, but, having done so, let us avow frankly that at bottom we know nothing; that we are imbedded in a mystery, towards the solution of which no whisper has been yet conceded to the listening intellect of man.

The world of life and beauty is now retreating, and the world of death and beauty is at hand. We were soon at the end of the Gauli glacier, from which the impetuous Urbach rushes, and turned into a chalet for a draught of milk. The Senner within proved an extortioner—' *ein unverschämter Hund*;' but let him pass. We worked along the flank of the glacier to a point which commands a view of the cliffy barrier which it is the main object of our journey to pass. From a range of snow-peaks linked together by ridges of black rock, the Gauli glacier falls, at first steeply as snow, then more gently as ice. We scan the mountain barrier to ascertain where it ought to be attacked. No one of us has ever been here before, and the scanty scraps of information which we have received tell us that at one place only is the barrier passable. We may reach the summit at several points from this side, but all save one, we are informed, lead to the brink of intractable precipices, which fall sheer to the Lauteraar glacier. We observe, discuss, and finally decide. We enter upon the glacier; black chasms yawn here and there through the super-incumbent snow, but there is no real difficulty. We cross the glacier and reach the opposite slopes; our way first lies up a moraine, and afterwards through the snow; a laborious ascent brings us close to the ridge, and here we pause once more in consultation. There is a gentle indentation to our left, and a cleft in the rocks to our right; our information points to the cleft, but we decide in favour of the saddle.

The winter snows were here thickly laid against the precipitous crags; the lower part of the buttress thus formed had broken away from the upper, which still clung to the rocks, the whole ridge being thus defended by a profound chasm, called in Switzerland a Bergschrund. At some places portions of snow had fallen away from the upper slope and partially choked the schrund, closing, however, its mouth only, and on this snow we were now to seek a footing. Bennen and myself were loose coming up; Forster and his guide were tied together; but now my friend declares that we must all be attached. We accordingly rope ourselves, and advance along the edge of the fissure to a place where it is partially stopped. A vertical wall of snow faces us. Our leader carefully treads down the covering of the chasm; and having thus rendered it sufficiently rigid to stand upon, he cuts a deep gap with his ice-axe in the opposing wall. Into the gap he tries to force himself, but the mass yields, and he falls back, sinking deeply in the snow of the schrund. He stands right over the fissure, which is merely bridged by the snow. I call out, 'Take care!' he responds, 'All right!' and returns to the charge. He hews a deeper and more ample gap; strikes his axe into the slope above him, and leaves it there;

buries his hands in the yielding mass, and raises his
body on his two arms, as on a pair of pillars. He
thus clears the schrund, and anchors his limbs in
the snow above. I am speedily at his side, and we
both tighten the rope as our friend Forster advances.
With perfect courage and a faultless head, he has
but one disadvantage, and that is an excess of
weight of at least two stone. In his first attempt
the snow-ledge breaks, and he falls back; but two
men are now at the rope, the tension of which,
aided by his own activity, prevents him from sinking far. By a second effort he clears the difficulty,
is followed by his guide, and all four of us reach
the slope above the chasm. Its steepness was
greater than that of a cathedral roof, while below
us, and within a few yards of us, was a chasm into
which it would be certain death to fall. Education
enables us to regard a position of this kind almost
with indifference; still the work was by no means
unexciting. In this early stage of our summer
performances, it required perfect trust in our leader
to keep our minds at ease. We reached the saddle,
and a cheer at the summit announced that our
escape was secured.

The indentation formed the top of a kind of
chimney or funnel in the rocks, which led right down
to the Lauteraar glacier. Elated with our success,
I released myself from the rope and sprang down

the chimney, preventing the descent from quicken-
ing to an absolute fall by seizing at intervals the
projecting rocks.   Once an effort of this kind shook
the alpenstock from my hand ; it slid along the
rubbish, reached a snow-slope, shot down it, and was
caught on some shingle at the bottom of the slope.
Quickly skirting the snow, which, without a staff,
cannot be trusted, I reached a ridge, from which a
jump landed me on the *débris* : it yielded and carried
me down ; passing the alpenstock I seized it, and in
an instant was master of all my motions.   Another
snow-slope was reached, down which I shot to the
rocks at the bottom, and there awaited the arrival of
my guide.

We diverged from the deep cut of the chimney,
Bennen adhering to the rough rocks, while I, hoping
to make an easier descent through the funnel itself,
resorted to it.   It was partially filled with indurated
snow, but underneath was a stream, and my ignorance of the thickness of the roof rendered caution
necessary.   At one place the snow was broken quite
across, and a dark tunnel, through which the stream
rushed, opened immediately below me.   My descent
being thus cut off, I crossed the couloir to the opposite
rocks, climbed them, and found myself upon the summit of a ledged precipice, below which Bennen stood,
watching me as I descended.   On one of the ledges
my foot slipped ; a most melancholy whine issued

from my guide, as he suddenly moved towards me; but the slip in no way compromised me; I reached the next ledge, and in a moment was clear of the difficulty. We dropped down the mountain together, quitted the rocks, and reached the glacier, where we were soon joined by Forster and his companion. Turning round, we espied a herd of seven chamois on one of the distant slopes of snow. The telescope reduced them to five full-grown animals and two pretty little kids. The day was fading and the deeper glacier pools were shaded by their icy banks. Through the shadowed water needles of ice were darting: all day long the molecules had been kept asunder by the antagonistic heat; their enemy is now withdrawn, and they lock themselves together in a crystalline embrace. Through a reach of merciless shingle, which covers the lower part of the glacier, we worked our way; then over green pastures and rounded rocks, to the Grimsel Hotel, which, uncomfortable as it is, was reached with pleasure by us all.

VII.

THE GRIMSEL AND THE ÆGGISCHHORN.

THIS Grimsel is a weird region—a monument carved with hieroglyphics more ancient and more grand than those of Nineveh or the Nile. It is a world disinterred by the sun from a sepulchre of ice. All around are evidences of the existence and the might of the glaciers which once held possession of the place. All around the rocks are carved, and fluted, and polished, and scored. Here and there angular pieces of quartz, held fast by the ice, inserted their edges into the rocks and scratched them like diamonds, the scratches varying in depth and width according to the magnitude of the cutting stone. Larger masses, held similarly captive, scooped longitudinal depressions in the rocks over which they passed, while in many cases the polishing must have been effected by the ice itself. A raindrop will wear a stone away; much more would an ice surface, squeezed into perfect contact by enormous pressure, rub away the asperities of the rocks over which for

ages it was forced to slide.   The rocks thus polished
by the ice itself are so exceedingly smooth and
slippery that it is impossible to stand on them where
their inclination is at all considerable.   But what a
world it must have been when the valleys were thus
filled! We can restore the state of things in
thought, and in doing so we submerge many a mass
which now lifts its pinnacle skyward.   Switzerland
in those days could not be so grand as it is now.
Pour ice into those valleys till they are filled, and
you eliminate those contrasts of height and depth
on which the grandeur of Alpine scenery depends.
Instead of skiey pinnacles and deep-cut gorges we
should have an icy sea dotted with dreary islands
formed by the highest mountain-tops.

In the afternoon I strolled up to the Siedelhorn.
As I stood upon the broken summit of the mountain
the air was without a cloud ; and the sunbeams fell
directly against the crown and slopes of the
Galenstock at the base of which lay the glacier of
the Rhone.   The level sea of *névé* above the great
ice-cascade, the fall itself, and the terminal glacier
below the fall were all apparently at hand.   At the
base of the fall the ice undergoes an extraordinary
transformation ; it reaches this place more or less
amorphous, it quits it most beautifully laminated,
the change being due to the pressure endured at
the bottom of the fall.   The wrinkling of the

glacier here was quite visible, the dwindling of the wrinkles into bands, and the subdivision of these bands into lines which mark the edges of the laminæ of which the glacier at this place is made up. Beyond, amid the mountains at the opposite side of the Rhone valley, lay the Gries glacier, half its snow in shadow, and half illuminated by the sinking sun. Round farther to the right were the Monte Leone and other grand masses, the grandest here being the Mischabel with its crowd of snowy cones. Jumping a gap in the mountains, we hit the stupendous cone of the Weisshorn, which slopes to meet the inclines of the Mischabel, and in the wedge of space carved out between the two the Matterhorn lifts its terrible head.

Wheeling farther in the same direction, we at length strike the mighty spurs of the Finsteraarhorn, between two of which lies the Oberaar glacier. Here is no turmoil of crevasses, no fantastic ice-pinnacles, nothing to indicate the operation of those tremendous forces by which a glacier sometimes rends its own breast. The grimmest giant of the Oberland closes the view at the head of the Lauteraar glacier—the Schreckhorn, whose cliffs on this side no mountaineer will ever scale. Between the Schreckhorn and Finsteraarhorn a curious group of peaks encircle a flat snow-field, from which the sunbeams are flung in blazing lines. Immediately below is the Unteraar glacier, with a long black streak upon its back, bent hither and thither, like a serpent wriggling down the valley. Beyond it and flanking it is a ridge of mountains with a crest of vertical rock, hacked into indentations which suggest a resemblance to a cock's comb. To the very root of the comb the mountains have been planed by the ancient ice.

A scene of unspeakable desolation it must have been when not Switzerland alone, but all Europe, was thus encased in frozen armour—when a glacier from Ben Nevis dammed the mouth of Glenroy, and Llanberis and Borrodale were ploughed by frozen shares sent down by Snowdon and Scawfell—when from the Reeks of Magillicuddy came the navigators which dug out space for the Killarney lakes, and carved through the mountains the Gap of Dunloe.[1] Evening came, and I moved downwards, over heaped boulders and tufted alp; down with headlong speed over the rounded rocks of the Grimsel, making long springs at intervals, over the polished inclines, and reaching the hospice as the bell rings its inmates to their evening meal.

On Saturday I ascended from Viesch to the Hotel Jungfrau on the slope of the Æggischhorn, and in the evening walked up to the summit of the mountain alone. As usual, I wandered unconsciously

[1] See chapter on 'Killarney,' p. 413.

from the beaten track, getting into a chaos of crags which had been shaken from the heights. My ascent was quick, and I soon found myself upon the crest of broken rocks which caps the mountain. The peak and those adjacent, which are similarly shattered, exhibit a striking picture of the ruin which nature inflicts upon her own creations. She buildeth up and taketh down. She lifts the mountains by her subterranean energies, and then blasts them by her lightnings and her frost. Thus grandly she rushes along the ' grooves of change ' to her unattainable repose. Is it unattainable ? The incessant tendency of material forces is toward final equilibrium ; and if the quantity of this tendency be finite, a time of repose must come at last. If one portion of the universe be hotter than another, a flux instantly sets in to equalise the temperatures; while winds blow and rivers roll in search of a stable equilibrium. Matter longs for rest ; when is this longing to be fully satisfied ? If satisfied, what then ? Rest is not perfection ; it is death. Life is only compatible with mutation ; when equilibrium sets in life ceases, and the world thenceforward is locked in everlasting sleep.

A wooden cross bleached by many storms surmounts the pinnacle of the Æggischhorn, and at the base of it I now take my place and scan the surrounding scene. Down from its birthplace in

the mountains comes that noblest of ice-streams the
Great Aletsch glacier. Its arms are thrown round
the shoulders of the Jungfrau, while from the Monk
and the Trugberg, the Gletscherhorn, the Breithorn,
the Aletschhorn, and many another noble pile, the
tributary snows descend and thicken into ice. The
mountains are well protected by their wintry coats,
and hence the quantity of *débris* upon the glacier is
comparatively small; still, along it can be noticed
dark longitudinal streaks, which are incipient
moraines. Right and left from these longitudinal
bands sweep finer curves, twisted here and there
into complex windings, which mark the lamination
of the subjacent ice. The glacier lies in a curved
valley, the side towards which its convex curvature
is turned is thrown into a state of strain, the ice
breaks across the line of tension, a curious system of
oblique glacier ravines being thus produced. From
the snow-line which crosses the glacier above the
Faulberg a pure snow-field stretches upward to the
Col de la Jungfrau, which unites the Maiden to the
Monk. Skies and summits are to-day without a
cloud, and no mist or turbidity interferes with the
sharpness of the outlines. Jungfrau, Monk, Eiger,
Trugberg, cliffy Strahlgrat, stately, lady-like Aletschhorn, all grandly pierce the empyrean. Like a Saul of
mountains, the Finsteraarhorn overtops all his neighbours; then we have the Oberaarhorn, with the riven

glacier of Viesch rolling from his shoulders. Below is the Märjelin See, with its crystal precipices and its floating icebergs, snowy white, sailing on a blue-green sea. Beyond is the range which divides the Valais from Italy. Sweeping round, the vision meets an aggregate of peaks which look, as fledglings to their mother, towards the mighty Dom. Then come the repellent crags of Mont Cervin, the idea of moral savagery, of wild untameable ferocity, mingling involuntarily with our contemplation of the gloomy pile. Next comes an object scarcely less grand, conveying it may be even a deeper impression of majesty and might than the Matterhorn itself— the Weisshorn, perhaps the most splendid object in the Alps. But beauty is associated with its force, and we think of it, not as cruel, but as grand and strong. Further to the right is the Great Combin; other peaks crowd around him, while at the extremity of the curve along which the gaze has swept rises the sovran crown of Mont Blanc. And now, as the day sinks, scrolls of pearly clouds form around the mountain-crests, and are wafted from them into the distant air. They are without colour of any kind; but their grace of form and lustre are not to be described.

NOTE ON CLOUDS.

IT is well known that when a receiver filled with ordinary undried air is exhausted, a cloudiness, due to the precipitation of the aqueous vapour diffused in the air, is produced by the first few strokes of the pump. It is, as might be expected, possible to produce clouds in this way with the vapours of other liquids than water.

In the course of some experiments on the chemical action of light on vapours which have been communicated to the Royal Society, I had frequent occasion to observe the precipitation of such clouds; indeed, several days at a time have been devoted solely to the generation and examination of clouds formed by the sudden dilatation of mixed air and vapours in the experimental tubes.

The clouds were generated in two ways: one mode consisted in opening the passage between the filled experimental tube and the air-pump, and then simply dilating the air by working the pump. In the other, the experimental tube was connected with a vessel of suitable size, the passage between which and the experimental tube could be closed by a stopcock. This vessel was first exhausted; on turning the cock the air rushed from the experimental tube into the vessel, the precipitation of a cloud within the tube being a consequence of the transfer. Instead of a special vessel, the cylinders of the air-pump itself were usually employed for this purpose.

It was found possible, by shutting off the residue of air and vapour after each act of precipitation, and again exhausting the cylinders of the pump, to obtain with some substances, and without refilling the experimental tube, fifteen or twenty clouds in succession.

The clouds thus precipitated differed from each other in luminous energy, some shedding forth a mild white light, others flashing out with sudden and surprising brilliancy. This difference of action is, of course, to be referred to the different reflective energies of the particles of the clouds, which were produced by substances of very different refractive indices.

Different clouds, moreover, possess very different degrees of stability : some melt away rapidly, while others linger for minutes in the experimental tube, resting, as they slowly dissolve, upon its bottom like a heap of snow. The particles of other clouds are trailed through the experimental tube as if they were moving through a viscous medium.

Nothing can exceed the splendour of the diffraction phenomena exhibited by some of these clouds; the colours are best seen by looking along the experimental tube from a point above it, the face being turned towards the source of illumination. The differential motions introduced by friction against the interior surface of the tube often cause the colours to arrange themselves in distinct layers.

The difference in texture exhibited by different clouds caused me to look a little more closely than I had previously done into the mechanism of cloud-formation. A certain expansion is necessary to bring down the cloud; the moment just before precipitation the cooling air and vapour may be regarded as divided into a number of polyhedra, the particles along the bounding surfaces of which move in opposite directions when precipitation actually sets in. Every cloud-particle has consumed a polyhedron of vapour in its formation; and it is manifest that the size of the particle must depend, not only on the size of the vapour

polyhedron, but also on the relation of the density of the
vapour to that of its liquid. If the vapour were light, and
the liquid heavy, other things being equal, the cloud-
particle would be smaller than if the vapour were heavy
and the liquid light. There would evidently be more
shrinkage in the one case than in the other; these considerations were found valid throughout the experiments;
the case of toluol may be taken as representative of a great
number of others. The specific gravity of this liquid is
0·85, that of water being unity; the specific gravity of its
vapour is 3·26, that of aqueous vapour being 0·6. Now,
as the size of the cloud-particle is directly proportional to
the specific gravity of the vapour, and inversely proportional to the specific gravity of the liquid, an easy calculation proves that, assuming the size of the vapour polyhedra
in both cases to be the same, the size of the particle of
toluol cloud must be more than six times that of the
particle of aqueous cloud. It is probably impossible
to test this question with numerical accuracy; but the
comparative coarseness of the toluol cloud is strikingly
manifest to the naked eye. The case is, as I have
said, representative.

In fact, aqueous vapour is without a parallel in these
particulars; it is not only the lightest of all vapours, in
the common acceptation of that term, but the lightest of
all gases except hydrogen and ammonia. To this circumstance the soft and tender beauty of the clouds of our
atmosphere is mainly to be ascribed.

The *sphericity* of the cloud-particles may be immediately
inferred from their deportment under the luminous beams.
The light which they shed when spherical is *continuous*:
but clouds may also be precipitated in solid flakes; and
then the incessant sparkling of the cloud shows that its

particles are *plates*, and not spheres. Some portions of the same cloud may be composed of spherical particles, others of flakes, the difference being at once manifested through the *calmness* of the one portion of the cloud, and the *uneasiness* of the other. The sparkling of such flakes reminded me of the plates of mica in the River Rhone at its entrance into the Lake of Geneva when shone upon by a strong sun.—*Proceedings of the Royal Society*, vol. xvii. p. 317.

Clouds are so often referred to in these pages that I thought it might be of interest to note the latest remarks on their formation.

VIII.

THE BEL ALP.

On Tuesday the 13th I accompanied a party of friends to the Märjelin See, skirted the lake, struck in upon the glacier, and having heard much of the position and the comfort of a new hotel upon the Bel Alp, I resolved to descend the glacier and pay the place a visit. The Valais range had been covered before we quitted the Æggischhorn; and, though the sun rode unimpeded in the higher heavens, vast masses of cloud continued to thrust themselves forth like tree-branches into the upper air.

The clouds extended, becoming ever blacker, until finally they were unlocked by thunder, and shook themselves down upon us in furious rain. The glacier is here cut up into oblique valleys of ice, subdivided by sharp-edged crevasses. We advanced swiftly along the ridges, but these finally abutted against the mountain, and we were compelled to cross from ridge to ridge. Hirst followed Bennen, and I trusted to my own devices. Joyously we struck our axes into the rumbling crests, and made our way rapidly between the chasms. The sunshine gushed down upon us, and partially dried our drenched clothes. At some distance to our left we observed upon the ice a group of persons, consisting of two men, a boy, and an old woman, engaged beside a crevasse; a thrill of horror shot through me, at the thought of a man being possibly between its jaws. We quickly joined them, and found an unfortunate cow firmly jammed between the frozen sides of the fissure, and groaning piteously. The men seemed very helpless; their means were inadequate, and their efforts ill-directed. ' Give the brute space, cut away the ice which presses the ribs, and *you* step upon that block which stops the chasm, and apply your shoulders to the creature's buttocks.' The ice splinters fly aloft, under the vigorous strokes of Bennen. Hirst suggests that a rope should be passed round the horns, so as to enable all hands to join in the pull. This is done. Another rope is passed between the hind legs. Bennen has loosened the ice which held the ribs in bondage, and now, like mariners heaving an anchor, we all join in a tug, timing our efforts by an appropriate exclamation. The weight moves, but extremely little; again the cry, and again the heave—it moves a little more. This is repeated several times till the fore-legs are extricated and thrown forward on the ice. We now lift the hinder parts, and succeed in placing the animal

upon the glacier, panting and trembling all over.
Folding our rope, we went onward. The day again
darkened. Again the thunder rang, being now preceded by lightning, which was thrown into my eyes
from the polished surface of my axe. Flash followed
flash and peal succeeded peal with terrific grandeur,
and the loaded clouds sent down from all their
fringes dusky streamers of rain. They looked like
waterspouts, so dense was their texture. Furious as
was the descending shower, hard as we were hit by the
mixed pellets of ice and water, I enjoyed the scene.
Grandly the cloud-besom swept the mountains, their
colossal outlines looming at intervals like over-
powered Titans struggling against their doom.

The glacier becoming impracticable through crevasses, we retreated to its eastern shore, and got along
the lateral moraine. It was rough work. The slope to
our left was partially clothed with spectral pines.
Storms had stripped the trunks of their branches,
and the branches of their leaves, leaving the tree-
wrecks behind, as if spirit-stricken and accursed.
Our home is now in sight, perched upon the summit
of a bluff opposite. We passed swiftly over the
ridges towards our destination. Wet and thirsty, we
reached the opposite side, and, striking into a beaten
track, finally reached the pleasant auberge at which
our journey ends.

From the hotel on the slope of Æggischhorn an

hour's ascent is required to place you in presence of
the magnificent view from the summit. But the
very windows of the hotel upon the Bel Alp command
noble views, and you may sit upon the bilberry
slopes adjacent before the grandest of mountain
scenes. On the 14th I went down to the savage
gorge in which the Aletsch glacier ends. A pine-
tree stood sheer over it ; bending its trunk at a right
angle near its roòt, and grasping a rock with its root,
it supported itself above the chasm. Standing upon
the horizontal part of the tree, I hugged its upright
stem, and looked down into the gorge. It required
several minutes to chase away my timidity, and
as the wind blew more forcibly against me, I clung
with greater fixity to the tree. In this wild spot,
and alone, I watched the dying fires of the day, until
the latest glow had vanished from the mountains.

Above the Bel Alp, and two hours distant, is
the grey pinnacle of the Sparrenhorn. I went up
there on the 15th. To the observer from the
hotel it appears as an isolated peak ; but it forms
the lofty end of a narrow ridge, which is torn
into ruins by the weather. At a distance in front of
me was a rocky promontory like the Abschwung,
right and left of which descended two streams of ice,
which welded themselves to a common trunk glacier.
The scene was perfectly unexpected and strikingly
beautiful. Nowhere have I seen more perfect repose,

nowhere more tender curves or finer structural lines. The stripes of the moraine bending along the glacier contribute to its beauty, and its deep seclusion gives it a peculiar charm. It seems a river so protected by its bounding mountains that no storm can ever reach it, and no billow disturb the perfect serenity of its rest. The sweep of the Aletsch glacier is also mighty as viewed from this point, and from no other could the Valais range seem more majestic. It is needless to say a word about the grandeur of the Dom, the Cervin, and the Weisshorn, all of which, and a great deal more, are commanded from the Sparrenhorn.

THE WEISSHORN FROM THE RIFFEL.

IX.

THE WEISSHORN.

ON Friday the 16th of August I rose at 4.30 ; the eastern heaven was hot with the glow of the rising sun, and against it were drawn the mountain outlines. At 5.30 I bade good-bye to the excellent little auberge of the Bel Alp,[1] and went straight down the mountain to Briegg, took the diligence to Visp, and engaged a porter immediately to Randa. I had sent Bennen thither to inspect the Weisshorn. On my arrival I learned that he had made the necessary reconnaissance, and entertained hopes of our being able to gain the top.

This noble mountain, which is fourteen thousand eight hundred feet high, had been tried on various occasions and from different sides by brave and competent climbers, but all efforts had been hitherto unavailing.

Previous to quitting Randa to assail this formidable peak I had two pairs of rugs sewed together so as to form two sacks. These and other coverlets,

---

[1] Now a substantial hotel which merits encouragement.

together with our wine and provisions, were sent on in advance of us. At 1 P.M. on the 18th of August Bennen, Wenger, and myself quitted the hotel, and were soon zigzagging among the pines of the opposite mountain. Wenger had been the guide of my friend Forster, and had shown himself so active and handy on the Strahleck that I commissioned Bennen to engage him. During the previous night I had been very unwell, and as I climbed the slope I suffered from intense thirst. Water seemed powerless to quench the desire for drink. We reached a chalet, and at our request a smart young Senner caught up a pail, and soon returned with it full of delicious milk. The effect of the milk was astonishing. It seemed to lubricate every atom of my body, and to exhilarate with its fragrance my brain.

Two hours' additional climbing brought us to our bivouac, a ledge of rock which jutted from the mountain-side, and formed an overhanging roof. On removing the stones from beneath the ledge, a space of comparatively dry clay was laid bare. This was to be my bed, and to soften it Wenger considerately stirred it up with his axe. The position was excellent, for lying upon my left side I commanded the whole range of Monte Rosa, from the Mischabel to the Breithorn. We were on the edge of an amphitheatre. Beyond the Schallenbach was the stately Mettelhorn. A row of eminences swept round to the right linked by lofty ridges of cliffs, which embraced the Schallenberg glacier. They formed, however, only a spur of the vaster Weisshorn, the cone of which was not visible from our dormitory. In company with Bennen I afterwards skirted the mountain until the whole colossal pyramid stood facing us. When I first looked at it my hopes sank, but both of us gathered confidence from a more lengthened gaze. The mountain is a pyramid with three faces, the intersections of which form three sharp edges or *arêtes*. The end of the eastern ridge was nearest to us, and on it our attention was principally fixed. We finally decided on the route to be pursued next morning, and with a chastened hope in both our breasts we returned to our shelter.

Water was our first necessity : it seemed everywhere, but there was none to drink. It was locked to solidity in the ice and snow. The sound of it came booming up from the Vispbach, as it broke into foam or rolled its boulders over its waterworn bed ; and the swish of many a minor streamlet mingled with the muffled roar of the large one. Bennen set out in search of the precious liquid, and after a long absence returned with a jug and pan full. At our meal, Wenger, who is a man rich in small expedients, turned the section of a cheese towards the flame of our pine fire ; it fizzed and blistered and turned viscous, and, the toasted surface being removed, was consumed

with relish by us all. The sunset had been unspeakably grand, steeping the zenith in violet, and
flooding the base of the heavens with crimson light.
Immediately opposite to us rose the Mischabel, with
its two great peaks, the Grubenhorn and the Täsch-
horn, each barely under 15,000 feet in height.
Next came the Alphubel, with its flattened crown of
snow; then the Allaleinhorn and Rympfischhorn;
then the Cima di Jazzi; next the mass of Monte
Rosa, flooded with light from bottom to top. The
face of the Lyskamm turned towards us was for the
most part shaded, but here and there its projecting
portions jutted forth red hot as the light fell upon
them The 'Twins' were most singularly illuminated; across the waist of each of them was drawn a
black bar, produced by the shadow of a corner of the
Breithorn, while their bases and crowns were exposed
to the crimson light. Over the rugged face of the
Breithorn itself the light fell as if in splashes, ig-
niting its glaciers and swathing its black crags in a
layer of transparent red. The Mettelhorn was cold,
so was the entire range governed by the Weisshorn,
while the glaciers they embraced lay grey and
ghastly in the twilight shade.

The sunlight lingered, while up the arch of the
opposite heavens the moon, within one day of being
full, seemed hastening to our aid. She finally appeared exactly behind the peak of the Rympfischhorn,

the cone of the mountain being projected for a short time as a triangle on the lunar disc. Only for a short time, however; the silver sphere soon cleared the mountain, and bore away through the tinted sky. The motion was quite visible, and resembled that of a vast balloon. As the day approached its end the scene assumed the most sublime aspect. All the lower portions of the mountains were deeply shaded, while the loftiest peaks, ranged upon a semicircle, were fully exposed to the sinking sun. They seemed pyramids of solid fire, while here and there long stretches of crimson light drawn over the higher snow-fields linked the summits together. An intensely illuminated geranium flower seems to swim in its own colour, which apparently surrounds the petals like a layer, and defeats by its lustre any attempt of the eye to seize upon the sharp outline of the leaves. A similar effect was here observed upon the mountains; the glory did not seem to come from them alone, but seemed also effluent from the air around them. As the evening advanced, the eastern heavens low down assumed a deep purple hue, above which, and blending with it by infinitesimal gradations, was a belt of red, and over this again zones of orange and violet. I walked round the corner of the mountain at sunset, and found the western sky glowing with a more transparent crimson than that which overspread the east. The crown of the Weisshorn was imbedded in this magnificent light. After sunset the purple of the east changed to a deep neutral tint, and against the faded red which spread above it the sun-forsaken mountains laid their cold and ghastly heads. The ruddy colour vanished more and more; the stars strengthened in lustre, until finally the moon and they held undisputed possession of the sky.

My face was turned towards the moon until it became so chilled that I was forced to protect it by a light handkerchief. The power of blinding the eyes is ascribed to the moonbeams, but the real mischief is that produced by radiation from the eyes into clear space, and the inflammation consequent upon the chill. As the cold increased I was fain to squeeze myself more and more underneath the ledge, so as to lessen the space of sky against which my body could radiate. Nothing could be more solemn than the night. Up from the valley came the low thunder of the Vispbach. Over the Dom flashed in succession the stars of Orion, until finally the entire constellation hung aloft. Higher up in heaven was the moon, and her beams as they fell upon the snow-fields and pyramids were sent back in silvery lustre by some, while others remained a dead white. These, as the earth twirled round, came duly in for their share of the glory. The Twins caught it at length and retained it long, shining with a pure spiritual radiance, while the moon continued above the hills.

At twelve o'clock I looked at my watch, and a
second time at 2 A.M. The moon was then just
touching the crest of the Schallenberg, and we were
threatened with the withdrawal of her light. This
soon occurred. We rose at 2¼ A.M., consumed our
coffee, and had to wait idly for the dawn. A faint
illumination at length overspread the sky, and with
this promise of the coming day we quitted our
bivouac at 3½ A.M. No cloud was to be seen; as far
as the weather was concerned we were sure to have
fair play. We rounded the shingly shoulder of the
mountain to the edge of a snow-field, but before
entering upon it I disburthened myself of my strong
shooting jacket, leaving it on the mountain-side.
The sunbeams and my own exertion would, I knew,
keep me only too warm during the day. We crossed
the snow, cut our way through a piece of entangled
glacier, reached the Bergschrund, and passed it
without a rope. We ascended the frozen snow of the
couloir by steps, but soon diverged from it to the
rocks at our right, and mounted them to the end of
the eastern *arête* of the mountain.

A snow saddle separated us from the higher rocks.
With our staff-pikes at one side of the saddle, we
pass by steps cut upon the other. We find the
rocks hewn into fantastic turrets and obelisks, while
the loose chips of this sculpture are strewn confusedly upon the ridge. Amid these we cautiously

H

pick our way, winding round the towers or scaling
them amain. The work was heavy from the first,
the bending, twisting, reaching, and drawing up
calling upon all the muscles of the frame. After
two hours of this work we halted, and, looking back,
saw two moving objects on the glacier below us.
At first we took them to be chamois, but they were
men. The leader carried an axe, and his companion
a knapsack and an alpenstock. They followed our
traces, losing them apparently now and then, and
waiting to recover them. Our expedition had put
Randa in a state of excitement, and some of its best
climbers had urged Bennen to take them with him.
This he did not deem necessary, and now here were
two of them determined to try the thing on their
own account, and perhaps to dispute with us the
honour of the enterprise. On this point, however,
our uneasiness was small.

Resuming our gymnastics, the rocky staircase led
us to the flat summit of a tower, where we found
ourselves cut off from a similar tower by a deep gap
bitten into the mountain. The rope was here our
refuge. Bennen coiled it round his waist; we let
him down along the surface of the rock, until he
fixed himself on a ledge, where he could lend me a
helping hand. I followed him, and Wenger followed
me. By a kind of screw motion we twisted ourselves
round the opposite tower, and reached the ridge

behind it. Work of this kind, however, is not to be performed by the day, and, with a view of sparing our strength, we quitted the ridge and endeavoured to get along the southern slope of the pyramid. The mountain was scarred by long couloirs, filled with clear hard ice. The cutting of steps across these couloirs proved to be so tedious and fatiguing that I urged Bennen to abandon them and try the ridge once more. We regained it and worked along it as before. Here and there upon the northern side the snow was folded over, and we worked slowly upward along the cornice snow. The ridge became gradually narrower, and the precipices on each side more sheer. We reached the end of one of its subdivisions, and found ourselves separated from the next rocks by a gap about twenty yards across. The ridge has here narrowed to a mere wall, which, however, as rock, would present no serious difficulty. But upon the wall of rock was placed a second wall of snow, which dwindled to a pure knife-edge at the top. It was white, of very fine grain, and a little moist. How to pass this snow catenary I knew not, for I did not think a human foot could trust itself upon so frail a support. Bennen's practical sagacity, however, came into play. He tried the snow by squeezing it with his foot, and to my astonishment began to cross it. Even after the pressure of his feet the space he had to stand on did not exceed a hand-breadth. I followed him, exactly as a boy walking along a horizontal pole, with toes turned outwards. Right and left the precipices were appalling. We reached the opposite rock, and an earnest smile rippled over Bennen's countenance as he turned towards me. He knew that he had done a daring thing, though not a presumptuous one. 'Had the snow,' he said, ' been less perfect, I should not have thought of attempting it; but I knew after I had set my foot upon the ridge that we might pass without fear.'

It is quite surprising what a number of things the simple observation made by Faraday in 1846 enables us to explain. Bennen's instinctive act is justified by theory. The snow was fine in grain, pure, and moist. When pressed, the attachments of its granules were innumerable, and their perfect cleanness enabled them to freeze together with a maximum energy. It was this freezing which gave the mass its sustaining power.

Two fragments of ordinary table ice brought carefully together freeze and cement themselves at their place of junction; or if two pieces floating in water be brought together, they instantly freeze, and by laying hold of either of them gently you can drag the other after it through the water. Imagine such points of attachment distributed in

great numbers through a mass of snow. The substance becomes thereby a semi-solid instead of a
mass of powder. My guide, however, unaided by
any theory, did a thing from which I should have
shrunk, though backed by all the theories in the
world.

After this we found the rocks on the ridge so
shaken that it required the greatest caution to
avoid bringing them down upon us. With all our
care, moreover, we sometimes dislodged vast masses,
which leaped upon the slope adjacent, loosened
others by their shock, these again others, until
finally a whole flight of them would escape, setting
the mountain in a roar as they whizzed and thundered along its side to the snow-fields 4,000 feet
below us. The day was hot, the work hard, and
our bodies were drained of their liquids as by a
Turkish bath. To make good our loss we halted
at intervals where the melted snow formed liquid
veins, and quenched our thirst. A bottle of champagne, poured sparingly into our goblets over a little
snow, furnished Wenger and myself with many a
refreshing draught. Bennen feared his eyes, and
would not touch champagne. We, however, did not
find halting good ; for at every pause the muscles
became set, and some minutes were necessary to
render them again elastic. But for both mind and
body the discipline was grand. There is scarcely a

position possible to a human being which, at one
time or another during the day, I was not forced to
assume. The fingers, wrist, and forearm were my
main reliance, and as a mechanical instrument the
human hand appeared to me this day to be a miracle
of constructive art.

For the most part the summit was hidden from
us, but on reaching the successive eminences it came
frequently into view. After three hours spent on
the *arête*—about five hours, that is, subsequent to
starting—we saw the summit over another minor
summit, which gave it an illusive proximity. ' You
have now good hopes,' I remarked, turning to Bennen.
'I do not allow myself to entertain the idea of
failure,' he replied. Well, six hours passed on the
ridge, each of which put in its inexorable claim to
the due amount of mechanical work; and at the end
of this time we found ourselves apparently no nearer
to the summit than when Bennen's hopes cropped out
in confidence. I looked anxiously at my guide as
he fixed his weary eyes upon the distant peak.
There was no confidence in his expression; still I
do not believe that either of us entertained for a
moment the thought of giving in. Wenger complained of his lungs, and Bennen counselled him
several times to remain behind; but this the Oberland man refused to do. At the commencement of
a day's work one often feels anxious, if not timid;

but when the work is very hard we become callous, and sometimes stupefied by the incessant knocking about. This was my case at present, and I kept watch lest my indifference should become careless-ness. I repeatedly supposed a case where a sudden effort might be required of me, and felt all through that I had a fair residue of strength to fall back upon should such a call be made. This conclusion was sometimes tested by a spurt; flinging myself sud-denly from rock to rock, I proved my condition by experiment instead of relying on surmise. An eminence in the ridge which cut off the view of the summit was now the object of our exertions. We reached it; but how hopelessly distant did the summit appear! Bennen laid his face upon his axe for a moment; a kind of sickly despair was in his eye as he turned to me, remarking, 'Lieher Herr, die Spitze ist noch sehr weit oben.'

Lest the desire to gratify me should urge him beyond the bounds of prudence, I told my guide that he must not persist on my account; that I should cheerfully return with him the moment he thought it no longer safe to proceed. He replied that, though weary, he felt quite sure of himself, and asked for some food. He had it, and a gulp of wine, which mightily refreshed him. Looking at the mountain with a firmer eye, he exclaimed, 'Herr! wir müssen ihn haben,' and his voice, as he spoke,

rung like steel within my heart. I thought of
Englishmen in battle, of the qualities which had
made them famous: it was mainly the quality of not
knowing when to yield—of fighting for duty even
after they had ceased to be animated by hope.
Such thoughts helped to lift me over the rocks.
Another eminence now fronted us, behind which,
how far we knew not, the summit lay. We scaled
this height, and above us, but clearly within reach,
a silvery pyramid projected itself against the blue
sky. I was assured ten times over by my companions
that it was the highest point before I ventured to
stake my faith upon the assertion. I feared that it
also might take rank with the illusions which had so
often beset our ascent, and I shrunk from the consequent moral shock. A huge prism of granite, or
granitic gneiss, terminated the *arête*, and from it
a knife-edge of pure white snow ran up to a little
point. We passed along the edge, reached that
point, and instantly swept with our eyes the whole
range of the horizon. We stood upon the crown of
the redoubtable Weisshorn.

The long-pent feelings of my two companions
found vent in a wild and reiterated cheer. Bennen
shook his arms in the air and shouted as a Valaisian,
while Wenger raised the shriller yell of the Oberland.
We looked downwards along the ridge, and far below,
perched on one of its crags, could discern the two

Randa men. Again and again the roar of triumph
was sent down to them. They had accomplished
but a small portion of the ridge, and soon after our
success they wended their way homewards. They
came, willing enough, no doubt, to publish our
failure had we failed; but we found out afterwards
that they had been equally strenuous in announcing
our success; they had seen us, they affirmed, like
three flies upon the summit of the mountain. Both
men had to endure a little persecution for the truth's
sake, for nobody in Randa would believe that the
Weisshorn could be scaled, and least of all by a man
who for two days previously had been the object of
Philomène the waitress's constant pity, on account of
the incompetence of his stomach to accept all that
she offered for its acceptance. The energy of conviction with which the men gave their evidence had,
however, proved conclusive to the most sceptical
before we arrived.

Bennen wished to leave some outward and visible
sign of our success on the summit. He deplored
having no suitable flag; but as a substitute for such
it was proposed that he should use the handle of one
of our axes as a flagstaff, and surmount it by a red
pocket-handkerchief. This was done, and for some
time subsequently the extempore banner was seen
flapping in the wind. To his extreme delight, it
was shown to Bennen himself three days afterwards

by my friend Mr. Francis Galton, from the Riffelberg hotel.

Every Swiss climber is acquainted with the Weisshorn. I have long regarded it as the noblest of
all the Alps, and most other travellers share this
opinion. The impression it produces is in some
measure due to the comparative isolation with
which it juts into the heavens. It is not masked
by other mountains, and all around the Alps its
final pyramid is in view. Conversely, the Weisshorn
commands a vast range of prospect. Neither
Bennen nor myself had ever seen anything at all
equal to it. The day, moreover, was perfect; not
a cloud was to be seen; and the gauzy haze of the
distant air, though sufficient to soften the outlines
and enhance the colouring of the mountains, was
far too thin to obscure them. Over the peaks and
through the valleys the sunbeams poured, unimpeded save by the mountains themselves, which
sent their shadows in bars of darkness through the
illuminated air. I had never before witnessed a
scene which affected me like this one. I opened
my note-book to make a few observations, but soon
relinquished the attempt. There was something
incongruous, if not profane, in allowing the scientific faculty to interfere where silent worship
seemed the 'reasonable service.'

We had been ten hours climbing from our

bivouac to the summit, and it was now necessary
that we should clear the mountain before the close
of day. Our muscles were loose and numbed, and,
unless extremely urged, declined all energetic tension : the thought of our success, however, ran like
a kind of wine through our fibres and helped us
down. We once fancied the descent would be
rapid, but it was far from it. As in ascending,
Bennen took the lead; he slowly cleared each crag,
paused till I joined him, I pausing till Wenger
joined me, and thus one or other of us was always
in motion. Our leader showed a preference for
the snow, while I held on to the rocks, where my
hands could assist my feet. Our muscles were
sorely tried by the twisting round the splintered
turrets of the *arête*, but a long, long stretch of the
ridge must be passed before we can venture to swerve
from it. We were roused from our stupefaction at
times by the roar of the stones which we loosed from
the ridge and sent leaping down the mountain.
Soon after recrossing the snow catenary already
mentioned we quitted the ridge to get obliquely
along the slope of the pyramid. The face of it
was scarred by couloirs, of which the deeper and
narrower ones were filled with ice, while the others
acted as highways for the rocks quarried by the
weathering above. Steps must be cut in the ice, but
the swing of the axe is very different now from what

it was in the morning. Bennen's blows descended
with the deliberateness of a man whose fire is half-
quenched; still they fell with sufficient power, and
the needful cavities were formed. We retraced our
morning steps over some of the ice-slopes. No word
of warning was uttered here as we ascended, but
now Bennen's admonitions were frequent and emphatic—'Take care not to slip.' I imagined, however, that even if a man slipped he would be able to
arrest his descent; but Bennen's response when I
stated this opinion was very prompt—'No! it would
be utterly impossible. If it were snow you might
do it, but it is pure ice, and if you fall you will
lose your senses before you can use your axe.' I
suppose he was right. At length we turned directly
downwards, and worked along one of the ridges
which lie in the line of steepest fall. We first
dropped cautiously from ledge to ledge. At one
place Bennen clung for a considerable time to a face
of rock, casting out feelers of leg and arm, and
desiring me to stand still. I did not understand
the difficulty, for the rock, though steep, was by no
means vertical. I fastened myself on to it, Bennen
being on a ledge below, waiting to receive me.
The spot on which he stood was a little rounded
protuberance sufficient to afford him footing, but
over which the slightest momentum would have
carried him. He knew this, and hence his caution.

Soon after this we quitted our ridge and dropped
into a couloir to the left of it. It was dark, and
damp with trickling water. Here we disencumbered
ourselves of the rope, and found our speed greatly
augmented. In some places the rocks were worn
to a powder, along which we shot by glissades. We
swerved again to the left, crossed a ridge, and got
into another and dryer couloir. The last one was
dangerous, as the water exerted a constant sapping
action upon the rocks. From our new position we
could hear the clatter of stones descending the
gulley we had just forsaken. Wenger, who had
brought up the rear during the day, is now sent to
the front; he has not Bennen's power, but his legs
are long and his descent rapid. He scents out the
way, which becomes more and more difficult. He
pauses, observes, dodges, but finally comes to a
dead stop on the summit of a precipice, which
sweeps like a rampart round the mountain. We
moved to the left, and after a long *détour* succeeded
in rounding the precipice.

Another half-hour brings us to the brow of a
second precipice, which is scooped out along its
centre so as to cause the brow to overhang. Chagrin
was in Bennen's face: he turned his eyes upwards,
and I feared mortally that he was about to propose
a reascent to the *arête*. It was very questionable
whether our muscles could have responded to such a

demand. While we stood pondering here, a deep and confused roar attracted our attention. From a point near the summit of the Weisshorn, a rock had been discharged down a dry couloir, raising a cloud of dust at each bump against the mountain. A hundred similar ones were immediately in motion, while the spaces between the larger masses were filled by an innumerable flight of smaller stones. Each of them shook its quantum of dust in the air, until finally the avalanche was enveloped in a cloud. The clatter was stunning, for the collisions were incessant. Black masses of rock emerged here and there from the cloud, and sped through the air like flying fiends. Their motion was not one of translation merely, but they whizzed and vibrated in their flight as if urged by wings. The echoes resounded from side to side, from the Schallenberg to the Weisshorn and back, until finally, after many a deep-sounding thud in the snow, the whole troop came to rest at the bottom of the mountain. This stone avalanche was one of the most extraordinary things I had ever witnessed, and in connection with it I would draw the attention of future climbers of the Weisshorn to the danger which would infallibly beset any attempt to ascend it from this side, except by one of its *arêtes*. At any moment the mountain-side may be raked by a fire as deadly as that of cannon.

After due deliberation we moved along the precipice westward, I fearing that each step forward but
plunged us into deeper difficulty. At one place,
however, the precipice bevelled off to a steep incline of smooth rock, along which ran a crack, wide
enough to admit the fingers, and sloping obliquely
down to the lower glacier. Each in succession
gripped the rock and shifted his body sideways along
the crack until he came near enough to the glacier
to reach it by a rough glissade. We passed swiftly
along the glacier, sometimes running, and, on
steeper slopes, sliding, until we were pulled up for the
third time by a precipice which seemed even worse
than either of the others. It was quite sheer, and
as far as I could see right or left altogether hopeless.
To my surprise, both the men turned without hesitation to the right. I felt desperately blank, but I
could notice no expression of dismay in the countenance of either of my companions. They inspected
the moraine matter over which we walked, and at
length one of them exclaimed, ' Da sind die Spuren,'
lengthening his strides at the same moment. We
looked over the brink at intervals, and at length
discovered what appeared to be a mere streak of
clay on the face of the precipice. On this streak
we found footing. It was by no means easy, but to
hard-pushed men it was a deliverance. The streak
vanished, and we must get down the rock. This

fortunately was rough, so that by pressing the hands
against its rounded protuberances, and sticking the
boot-nails against its projecting crystals, we let
ourselves gradually down. A deep cleft separated
the glacier from the precipice ; this was crossed, and
we were free, being clearly placed beyond the last
bastion of the mountain.

In this admirable fashion did my guides behave
on this occasion. The day previous to my arrival
at Randa they had been up the mountain, and they
then observed a solitary chamois moving along the
base of this very precipice, and making ineffectual
attempts to get up it. At one place the creature
succeeded; this spot they fixed in their memories,
and when they reached the top of the precipice they
sought for the traces of the chamois, found them,
and were guided by them to the only place where
escape in any reasonable time was possible. Our
way was now clear ; over the glacier we cheerfully
marched, escaping from the ice just as the moon
and the eastern sky contributed about equally to the
illumination. The moonlight was afterwards intercepted by clouds. In the gloom we were often at a
loss, and wandered half-bewildered over the grassy
slopes. At length the welcome tinkle of cow-bells
was heard in the distance, and guided by them we
reached the chalet a little after 9 P.M. The cows
had been milked and the milk disposed of, but the

men managed to get us a moderate draught.   Thus
refreshed we continued the descent.   I was half
famished, for my solid nutriment during the day
consisted solely of part of a box of meat lozenges
given to me by Mr. Hawkins.   Bennen and myself
descended the mountain deliberately, and after many
windings emerged upon the valley, and reached the
hotel a little before 11 P.M.   I had a basin of broth,
not made according to Liebig, and a piece of mutton
boiled probably for the fifth time.   Fortified by
these, and comforted by a warm footbath, I went
to bed, where six hours' sound sleep chased away all
consciousness of fatigue.   I was astonished on the
morrow to find the loose atoms of my body knitted
so firmly by so brief a rest.   Up to my attempt upon
the Weisshorn I had felt more or less dilapidated,
but here all weakness ended, and during my subsequent stay in Switzerland I was unacquainted with
infirmity.

I

X.

INSPECTION OF THE MATTERHORN.

ON the afternoon of the 20th we quitted Randa, with a threatening sky overhead. The considerate Philomène compelled us to take an umbrella, which we soon found useful. The flood-gates of heaven were unlocked, while, defended by our cotton canopy, Bennen and myself walked arm-in-arm to Zermatt. I instantly found myself in the midst of a circle of pleasant friends, some of whom had just returned from a successful attempt upon the Lyskamm. On the 22nd quite a crowd of travellers crossed the Theodule Pass; and, knowing that every corner of the hotel at Breuil would be taken up, I halted a day, so as to allow the people to disperse. Breuil commands a view of the southern side of the Matterhorn; and it was now an object with me to discover, if possible, upon the true peak of this formidable mountain, some ledge or cranny where three men might spend a night. The mountain may be accessible or inaccessible, but one thing seems certain, that starting from Breuil, or even from the chalets above Breuil, the work of reaching the summit is too much for a single day. But could a shelter be found amid the wild battlements of the peak itself, which would enable one to attack the obelisk at day-dawn, the possibility of conquest was so far an open question as to tempt a trial. I therefore sent Bennen on to reconnoitre, purposing myself to cross the Theodule alone on the following day.

On the afternoon of the 22nd I sauntered slowly up to the Riffel, leaning at times on the head of my axe, or sitting down upon the grassy' knolls, as my mood prompted. The air which filled the valleys of the Oberland, and swathed in mitigated density the highest peaks, was slightly opalescent, though still transparent, the floating particles forming so many *points d'appui*, from which the light was scattered through surrounding space. The whole medium glowed as if shone upon by a distant furnace, and through it the outline of the mountains loomed. The glow augmented as the sun sank, reached its maximun, paused, and then ran speedily down to a cold and colourless twilight.

Next morning at nine o'clock, with some scraps of information from the guides to help me on my way, I quitted the Riffel to cross the Theodule. I was soon followed by the domestic of the hotel. Bennen had requested him to see me to the edge of the glacier, and he now joined me with this intention.

He knew my designs upon the Matterhorn, and strongly deprecated them. ' Only think, Herr,' he urged, ' what will avail your ascent of the Weisshorn if you are smashed upon the Mont Cervin ? Mein Herr !' he added with condensed emphasis, ' thun Sie es nicht.' The whole conversation was in fact a homily, the strong point of which was the utter uselessness of success on the one mountain if it were to be followed by annihilation on the other. We reached the ridge above the glacier, where, handing him a trinkgeld, which I had to force on his acceptance, I bade him good-bye, assuring him that I would submit in all things to Bennen's opinion. He had the highest idea of Bennen's wisdom, and hence the assurance sent him home comforted.

I was soon upon the ice, once more alone, as I delight to be at times. As a habit going alone is to be deprecated, but sparingly indulged in it is a great luxury. There are no doubt moods when the mother is glad to get rid of her offspring, the wife of her husband, the lover of his mistress, and when it is not well to keep them together. And so, at rare intervals, it is good for the soul to feel the full influence of that ' society where none intrudes.' When the work is clearly within your power, when long practice has enabled you to trust your own eye and judgment in unravelling crevasses, and your

THE MATTERHORN.

own axe and arm in subduing their more serious
difficulties, it is an entirely new experience to be
alone amid those sublime scenes. The peaks wear
a more solemn aspect, the sun shines with a more
effectual fire, the blue of heaven is more deep and
awful, and the hard heart of man is often made as
tender as a child's. You contract a closer friendship for the universe in virtue of your more intimate contact with its parts. The glacier to-day filled
the air with low murmurs, while the sound of the
distant moulins rose to a kind of roar. The *débris*
rustled on the moraines, the smaller rivulets babbled in their channels, as they ran to join their
trunk, and the surface of the glacier creaked au-
dibly as it yielded to the sun. It seemed to breathe
and whisper like a living thing. To my left was
Monte Rosa and her royal court, to my right the
mystic pin-acle of the Matterhorn, which from a
certain point here upon the glacier attains its maximum sharpness. It drew my eyes towards it with
irresistible fascination as it shimmered in the blue,
too preoccupied with heaven to think even with
contempt on the designs of a son of earth to reach
its inviolate crest.

I crossed the Görner glacier quite as speedily as
if I had been professionally led. Then up the
undulating slope of the Theodule glacier, with a
rocky ridge to the right, over which I was informed

a rude track led to the pass of St. Theodule. I am not great at finding tracks, and I missed this one, ascending until it became evident that I had gone, too far. Near its higher extremity the crest of the ridge is cut across by three curious chasms, and one of these I thought would be a likely gateway through the ridge. I climbed the steep buttress of the spur and was soon in the fissure. Huge masses of rock were jammed into it, the presence of which gave variety to the exertion, calling forth strength, but not exciting fear. From the summit the rocks sloped gently down to the snow, and in a few minutes the presence of broken bottles on the moraine showed me that I had hit upon the track. Upwards of twenty unhappy bees staggered against me on the way: tempted by the sun, or wafted by the wind, they had quitted the flowery Alps to meet torpor and death in the ice-world above. From the summit I went swiftly down to Breuil, where I was welcomed by the host, welcomed by the waiter; loud were the expressions of content at my arrival; and I was informed that Bennen had started early in the morning to 'promenade himself' around the Matterhorn.

I lay long upon the Alp, scanning crag and snow in search of my guide. From the admirable account of the first attempt on the Matterhorn, drawn up by Mr. Hawkins,[1] it may be inferred that the

---

[1] Chapter III. of this volume.

ascent is not likely to be a matter of mere amuse-
ment. The account narrates that after climbing
for several hours in the face of novel difficulties,
my companion thought it wise to halt so as to
secure our retreat. I will here state in a few
words what occurred after our separation from
him. Bennen and I had first a hard scramble
up some very steep rocks, our motions giving to
those below us the impression that we were urging
up bales of goods instead of the simple weight of our
own bodies. Turning the corner of the ridge, we had
to cross an unpleasant slope of smooth rock, covered
by about eighteen inches of snow. In ascending,
this place was passed in silence, but in coming down
the fear arose that the superficial layer might slip
away with us. Bennen seldom warns me, but he
did so here emphatically, declaring his own power-
lessness to render any help should the footing give
way. Having crossed this slope in our ascent, we
were fronted by a cliff, against which we rose mainly
by aid of the felspar crystals protuberant from its
face. Midway up the cliff Bennen asked me to hold
on, as he did not feel sure that it was the best route.
I accordingly ceased moving, and lay against the rock
with legs and arms outstretched. Bennen climbed
to the top of the cliff, but returned immediately with
a flush of confidence in his eye. ' I will lead you to
the top,' he said excitedly. Had I been free I should

have cried 'Bravo!' but in my position I did not
care to risk the muscular motion which a hearty
bravo would demand.

Aided by the rope, I was at his side in a minute,
and we soon learned that his confidence was pre-
mature. Difficulties thickened round us; on no
other mountain are they so thick, and each of them
is attended by possibilities of the most blood-
chilling kind. Our mode of motion was this:
Bennen advanced while I held on to a rock, prepared for the jerk if he should slip. When he
had secured himself, he called out, 'Ich bin fest,
kommen Sie.' I then worked forward, sometimes
halting where he had halted, sometimes passing him
until a firm anchorage was gained, when it again
became his turn to advance. Thus each of us waited
until the other could seize upon something capable
of bearing the shock of a falling man. At some
places Bennen deemed a little extra assurance necessary; and here he emphasised his statement that
he was 'fest' by a suitable hyperbole. 'Ich bin fest
wie ein Mauer,—fest wie ein Berg, ich halte Sie
gewiss,' or some such expression.

Looking from Breuil, a series of moderate-sized
prominences are seen along the *arête* of the Matterhorn; but when you are near them, these black
eminences rise like tremendous castles in the air,
so wild and high as almost to quell all hope of

scaling or getting round them. At the base of one
of these edifices Bennen paused and looked closely
at the grand mass; he wiped his forehead, and
turning to me said, 'Was denken Sie, Herr?'—
'Shall we go on, or shall we return? I will do
what you wish.' 'I am without a wish, Bennen,' I
replied: 'where you go I follow, be it up or down.'
He disliked the idea of giving in, and would willingly have thrown the onus of stopping upon me.
We attacked the castle, and by a hard effort reached
one of its mid ledges, whence we had plenty of
room to examine the remainder. We might certainly have continued the ascent beyond this place,
but Bennen paused here. To a minute of talk succeeded a minute of silence, during which my guide
earnestly scanned the heights. He then turned
towards me, and the words seemed to fall from his
lips through a resisting medium, as he said, 'Ich
denke die Zeit ist zu kurz' (I think the time is too
short).

By this time each of the neighbouring peaks had
unfolded a cloud banner, remaining clear to wind-
ward, but having a streamer hooked on to its
summit and drawn far out into space by the moist
south wind. It was a grand and affecting sight,
grand intrinsically, but doubly impressive to feelings
already loosened by the awe inseparable from our
position. Looked at from Breuil, the mountain

shows two summits separated from each other by a possibly impassable cleft. Only the lower one of these could be seen from where we stood. I asked Bennen how high he supposed it to be above the point where we then stood; he estimated its height at 400 feet, I at 500 feet. Probably both of us were under the mark; however, I state the fact as it occurred. The object of my present visit to Brueil was to finish the piece of work thus abruptly broken off, and so I awaited Bennen's return with anxious interest.

At dusk I saw him striding down the Alp, and went out to meet him. I sought to gather his opinion from his eye before he spoke, but could make nothing out. It was perfectly firm, but might mean either pro or con. 'Herr,' he said at length, in a tone of unusual emphasis, ' I have examined the mountain carefully, and find it more difficult and dangerous than I had imagined. There is no place upon it where we could well pass the night. We might do so on yonder col upon the snow, but there we should be almost frozen to death, and totally unfit for the work of the next day. On the rocks there is no ledge or cranny which could give us proper harbourage; and starting from Breuil it is certainly impossible to reach the summit in a single day.' I was entirely taken back by this report. Bennen was evidently dead against any attempt upon the

mountain. 'We can, at all events, reach the lower
of the two summits,' I remarked. 'Even that is
difficult,' he replied; 'but when you have reached
it, what then ? The peak has neither name nor fame.'
I was silent; slightly irascible, perhaps; but it was
against my habit to utter a word of remonstrance or
persuasion. Bennen made his report with his eyes
open. He knew me well, and I think mutual trust
has rarely been more strongly developed between
guide and traveller than between him and me. I
knew that I had but to give the word and he would
face the mountain with me next day, but it would
have been inexcusable in me to deal thus with him.
So I stroked my beard, and, like Lelia in the ' Prin-
cess,' when

> Upon the sward
> She tapt her tiny silken-sandal'd foot,

I crushed the grass with my hobnails, seeking thus
a safety-valve for my disappointment.

My sleep was unsatisfying that night, and on the
following morning I felt a void within. The hope
of finishing my work creditably had been suddenly
dislodged, and, for a time, vacuity took its place.
It was like the removal of a pleasant drug or the
breaking down of a religious faith. I hardly knew
what to do with myself. One thing was certain—
the Italian valleys had no tonic strong enough to
set me right; the mountains alone could restore

what I had lost. Over the Joch then once more!
We packed up and bade farewell to the host and
waiter. Both men seemed smitten with a sudden
languor, and could hardly respond to my adieus.
They had expected us to be their guests for some
time, and were evidently disgusted at our want of
pluck. 'Mais, monsieur, il faut faire la pénitence
pour une nuit.' Veils of the silkiest cloud began to
draw themselves round the mountain, and to stretch
in long gauzy filaments through the air, where
they finally curdled up to common cloud, and lost
the grace and beauty of their infancy. Had they
condensed to thunder I should have been better
satisfied; but it was some consolation to see them
thicken so as to hide the mountain, and quench the
longing with which I should have viewed its unclouded head. The thought of spending some days
chamois-hunting occurred to me. Bennen seized the
idea with delight, promising me an excellent gun.
We crossed the summit, descended to Zermatt,
paused there to refresh ourselves, and went forward
to St. Nicholas, where we spent the night.

XI.

OVER THE MORO.

I HAD only seen one half of Monte Rosa; and from the Italian side the aspect of the mountain was unknown to me. I had been upon the Monte Moro three years ago, but looked from it merely into an infinite sea of haze. To complete my knowledge of the mountain it was necessary to go to Macugnaga, and over the Moro I accordingly resolved to go. But resolution had as yet taken no deep root, and on reaching Saas I was beset by the desire to cross the Alphubel. Bennen called me at three; but over the pass grey clouds were hanging, and, determined not to mar this fine excursion by choosing an imperfect day, I then gave it up. At seven o'clock, however, all trace of cloud had disappeared; it had been merely a local gathering of no importance, which the first sunbeams resolved into transparency. It was now, however, too late to think of the Alphubel, so I reverted to my original design, and at 9 A.M. started up the valley towards Mattmark. A party

of friends in advance contributed strongly to draw
me on in this direction.

Onward then we went through the soft green
meadows, with the river sounding to our right. The
sun showered gold upon the pines, and brought richly
out the colouring of the rocks. The blue wood-
smoke ascended from the hamlets, and the compa-
nionable grasshopper sang and chirruped right and
left. High up the sides of the mountains the rocks
were planed down to tablets by the ancient glaciers.
The valley narrowed, and we skirted a pile of
moraine-like matter, which was roped compactly together by the roots of the pines. Huge blocks here
choke the channel of the river, and raise its murmurs
to a roar. We emerge from shade into sunshine,
and observe the smoke of a distant cataract jetting
from the side of the mountain. Crags and boulders
are here heaped in confusion upon the hill-side, and
among them the hardy trees find a lodgment, asking
no nutriment from the stones—asking only a pedes-
tal on which they may plant their trunks and lift
their branches into the nourishing air. Then comes
the cataract itself, plunging in rhythmic gushes down
the shining rocks.

The valley again opens, and finds room for a little
hamlet—dingy hovels, with a white little church
in the midst of them ; patches of green meadow and
yellow rye, with the gleam of the river here and

there. The moon hangs over the Mischabelhörner, turning a face which ever waxes paler towards the sun. The valley in the distance seems shut in by the Allalein glacier, which is approached amid the waterworn boulders strewn by the river in its hours of turbulence. The rounded rocks are now beautified with lichens, and scattered trees glimmer among the heaps. Nature heals herself. She feeds the glacier and planes the mountains down. She fuses the glacier and exposes the dead rocks. But instantly her energies are exerted to neutralise the desolation, clothing the crags with beauty, and sending the wandering wind in melody through the branches of the pines.

At the Mattmark hotel, which stands at the foot of the Monte Moro, I was joined by a gentleman who had just liberated himself from an unpleasant guide. Bennen halted on the way to adjust his knapsack, while my companion and myself went on. We lost sight of my guide, lost the track also, and clambered over crag and snow to the summit, where we waited till Bennen arrived. The mass of Monte Rosa here grandly revealed itself from top to bottom. Dark cliffs and white snows were finely contrasted, and the longer I looked at it, the more noble and impressive did the mountain appear. We were very soon clear of the snow, and went straight down the declivity towards Macugnaga.

We put up at the Monte Moro, where a party of friends greeted me with a vociferous welcome. This was my first visit to Macugnaga, and, save as a caldron for the generation of fogs, I knew scarcely anything about it. But there were no fogs there at the time to which I refer, and the place wore quite a charmed aspect. I walked out alone in the evening, up through the meadows towards the base of Monte Rosa, and on no other occasion have I seen peace, beauty, and grandeur so harmoniously blended. Earth and air were exquisite, and I returned to the hotel brimful of content.

Monte Rosa with her peaks and spurs builds here a noble amphitheatre. From the heart of the mountain creeps the Macugnaga glacier. To the right a precipitous barrier extends to the Cima di Jazzi, and between the latter and Monte Rosa this barrier is scarred by two couloirs, one of which, or the cliff beside it, has the reputation of forming the old pass of the Weissthor. It had long been uncertain whether this so-called 'Alter Pass' had ever been used as such, and many superior mountaineers deemed it from inspection to be impracticable. All doubt on this point was removed this year; for Mr. Tuckett, led by Bennen, had crossed the barrier by the couloir most distant from Monte Rosa, and consequently nearest to the Cima di Jazzi. As I stood in front of the hotel in the

afternoon, I said to Bennen that I should like to
try the pass on the following day; in ten minutes
afterwards the plan of our expedition was arranged.
We were to start before the dawn, and, to leave
Bennen's hands free, a muscular young fellow named
Andermatten was engaged to carry our provisions.
It was also proposed to vary the proceedings by
assailing the ridge by the couloir nearest to Monte
Rosa.

XII.

THE OLD WEISSTHOR.

I was called by my host at a quarter before three.
The firmament of Monte Rosa was almost as black
as the,rocks beneath it, while above in the darkness
trembled the stars. At 4 A.M. we quitted the hotel.
We wound along the meadows, by the slumbering
houses, and the unslumbering river. The eastern
heaven soon brightened, and we could look direct
through the gloom of the valley at the opening of
the dawn. We threaded our way amid the boulders
which the torrent had scattered over the plain, and
among which groups of stately pines now find
anchorage. Some of the trees had exerted all their
force in a vertical direction, and rose straight, tall,
and mastlike, without lateral branches. We reached
a great moraine, grey with years, and clothed with
magnificent pines ; our way lay up it, and from the
top we dropped into a little dell of magical beauty.
Deep hidden by the glacier-built ridges, guarded by
noble trees, soft and green at the bottom, and tufted
round with bilberry bushes, through which peeped
here and there the lichen-covered crags, I have

rarely seen a spot in which I should so like to
dream away a day. Before I entered it, Monte
Rosa was still in shadow, but on my emergence
I noticed that her precipices were all aglow. The
purple colouring of the mountains observed on
looking down the valley was indescribable; out of
Italy I have never seen anything like it. Oxygen
and nitrogen could not produce the effect; some
effluence from the earth, some foreign constituent
of the atmosphere, developed in those deep valleys
by the southern sun, must sift the solar beams,
weaken the rays of medium refrangibility, and
blend the red and violet of the spectrum to that
incomparable hue. The air indeed is filled with
floating matters which vary from day to day, and
it is mainly to such extraneous substances that the
chromatic splendours of our atmosphere are to be
ascribed. The air south of the Alps is in this respect
different from that on the north, but a modicum
even of arsenic might be respired with satisfaction,
if warmed by the bloom which suffused the air of
Italy this glorious dawn.

The ancient moraines of the Macugnaga glacier
rank among the finest that I have seen; long, high
ridges tapering from base to edge, hoary with age,
but beautified by the shrubs and blossoms of to-
day. We crossed the ice and them. At the foot of
the old Weissthor lay couched a small glacier, which

had landed a multitude of boulders on the slope below it ; and amid these we were soon threading our way. We crossed the little glacier, which at one place strove to be disagreeable, and here I learned from the deportment of his axe the kind of work to which our porter had been previously accustomed. Half a dozen strokes shook the head of the implement from its handle. We reached the rocks to the right of the couloir and climbed them for some distance. At the base the ice was cut by profound fissures, which extended quite across, and rendered a direct advance up the gulley impossible ; but higher up we dropped down upon the snow.

Close to the rocks it was scarred by a furrow six or eight feet deep, and about twelve in width, evidently the track of avalanches, or of rocks let loose from the heights. Into this we descended. The bottom was firm, and roughened by stones which found a lodgment there. It seemed that we had here a very suitable roadway to the top. But a sudden crash was heard aloft. I looked upward, and right over the snow-brow which closed the view perceived a large brown boulder in the air, while a roar of unseen stones showed that the visible projectile was merely the first shot of a general cannonade. They appeared—pouring straight down upon us—the sides of the furrow preventing them from squandering their force in any other direction.

'Schnell!' shouted the man behind me, and there is a ring in the word, when sharply uttered in the Alps, that almost lifts a man off his feet. I sprang forward, but, urged by a sterner impulse, the man behind sprung right on to me. We cleared the furrow exactly as the first stone flew by, and once in safety we could calmly admire the energy with which the rattling boulders sped along.

Our way now lay up the couloir; the snow was steep, but knobbly, and hence but few steps were required to give the boots a hold. We crossed and recrossed obliquely, like a horse drawing a laden cart up hill. At times we paused and examined the heights. The view ended in the snow-fields above, but near the summit suddenly rose a high ice-wall. If we persisted in the couloir, this barrier would have to be surmounted, and the possibility of scaling it was very questionable. Our attention therefore was turned to the rocks at our right, and the thought of assailing them was several times mooted and discussed. They at length seduced us, and we resolved to abandon the snow. To reach the rocks, however, we had to recross the avalanche channel, which was here very deep. Bennen hewed a gap at the top of its flanking wall, and, stooping over, scooped steps out of its vertical face. He then made a deep hole, in which he anchored his left arm, let himself thus partly down, and with his right pushed the steps to the bottom. While this was going on small stones were continually flying down the gulley. Bennen reached the floor, and I followed. Our companion was still clinging to the snow-wall, when a horrible clatter was heard overhead. It was another stone avalanche, which there was hardly a hope of escaping. Happily a rock was here firmly stuck in the bed of the gulley, and I chanced to be beside it when the first huge missile appeared. This was the delinquent which had set the others loose. I was directly in the line of fire, but, ducking behind the boulder, I let the projectile shoot over my head. Behind it came a shoal of smaller fry, each of them, however, quite competent to crack a human life. 'Schnell!' with its metallic clang, rung from the throat of Bennen; and never before had I seen his axe so promptly and vigorously applied.

While this infernal cannonade was directed upon us we hung upon a slope of snow which had been pressed and polished to ice by the descending stones, and so steep that a single slip would have converted us into an avalanche also. Without steps of some kind we dared not set foot on the slope, and these had to be cut while the stone shower was falling on us. Mere scratches in the ice, however, were all the axe could accomplish, and on these we steadied ourselves with the energy of

desperate men. Bennen was first, and I followed
him, while the stones flew thick beside and between
us. My excellent guide thought of me more than
of himself, and once caught upon the handle of
his axe, as a cricketer catches a ball upon his bat,
a lump which might have finished my climbing.
The labour of his axe was here for a time divided
between the projectiles and the ice, while at every
pause in the volley 'he cut a step and sprang
forward.' Had the peril been less, it would have
been amusing to see our duckings and contortions
as we fenced with our swarming foes. A final jump
landed us on an embankment out of the direct line
of fire, and we thus escaped a danger extremely
exciting to us all.

We had next to descend an ice-slope to a place
at which the rocks could be invaded. Here
Andermatten slipped, shot down the slope, knocked
Bennen off his legs, but before the rope had jerked
me off mine the guide had stopped his flight. The
porter's hat, however, followed the rushing stones.
It was shaken off his head and lost. If discipline for
eye, limb, head, and heart be of any value, we had
it, and were still likely to have it, here. Our first
experience of the rocks was by no means comforting :
they were uniformly steep, and, as far as we could
judge from a long look upwards, they were likely to
continue so. A stiffer bit than ordinary intervened

now and then, making us feel how possible it was to be entirely cut off.

We at length reached real difficulty number one. All three of us were huddled together on a narrow ledge, with a smooth and vertical cliff above us. Bennen tried it in various ways, but he was several times forced back to the ledge. At length he managed to hook the fingers of one hand over the top of the cliff, while to aid his grip he tried to fasten his shoes against its face. But the nails scraped freely over the granular surface, and he had for a time to lift himself almost by a single arm. As he did so he had as ugly a place beneath him as a human body could well be suspended over. We were tied to him of course ; but the jerk, had his grip failed, would have been terrible. He raised at length his breast to a level with the top, and leaning over it he relieved the strain· Seizing upon something further on, he lifted himself quite to the top ; then tightened the rope, while I slowly worked myself over the face of the cliff after him. We were soon side by side, and immediately afterwards Andermatten, with his long unkempt hair, and face white with excitement, hung midway between heaven and earth supported by the rope alone. We hauled him up bodily, and as he stood upon the ledge his limbs quivered beneath him.

We now strained slowly upwards amid the maze

of crags, and scaled a second cliff, resembling, though in a modified form, that just described. There was no peace, no rest, no delivery from the anxiety 'which weighed upon the heart.' Bennen looked extremely blank, and often cast an eye downward to the couloir we had quitted, muttering aloud, 'Had we only held on to the snow!' He had soon reason to emphasize his ejaculation.

After climbing for some time, we reached a smooth vertical face of rock from which, right or left, there was no escape, and over which we must go. Bennen first tried it unaided, but was obliged to recoil. Without a lift of five or six feet the thing was impossible. When a boy I have often climbed a wall by placing a comrade in a stooping posture with his hands and head against the wall, getting on his back, and permitting him gradually to straighten himself till he became erect. This plan I now proposed to Bennen, offering to take him on my back. 'Nein, Herr!' he replied; 'nicht Sie, ich will es mit Andermatten versuchen.' I could not persuade him, so Andermatten got upon the ledge, and fixed his knee for Bennen to stand on. In this position my guide obtained a precarious grip, just sufficient to enable him to pass with safety from the knee to the shoulder. He paused here, and pulled away such splinters as might prove treacherous if he laid hold of them. He at length found a firm one, and had next to urge himself, not fairly upward, for right above us the top was entirely out of reach, but obliquely along the face of the cliff. He succeeded, anchored himself, and called upon me to advance.

The rope was tight, it is true, but it was not vertical, so that a slip would cause me to swing like a pendulum over the cliff's face. With considerable effort I managed to hand Bennen his axe, and while doing so my own staff escaped me and was irrecoverably lost. I ascended Andermatten's shoulders as Bennen did, but my body was not long enough to bridge the way to the guide's arm; so I had to risk the possibility of becoming a pendulum. A little protrusion gave my left foot some support. I suddenly raised myself a yard, and here was met by the iron grip of my guide. In a second I was safely stowed away in a neighbouring fissure. Andermatten now remained. He first detached himself from the rope, tied it round his coat and knapsack, which were drawn up. The rope was again let down, and the porter tied it firmly round his waist. It was not made in England, and was perhaps lighter than it ought to be; so to help it hands and feet were scraped with spasmodic energy over the rock. He struggled too much, and Bennen cried sharply, 'Langsam! langsam! Keine Furcht!' The poor fellow looked very pale and bewildered as his bare head emerged above the ledge. His body soon followed. Bennen always used the imperfect for the present tense, ' Er war ganz bleich,' he remarked to me, by the ' war' meaning *ist*.

The young man seemed to regard Bennen with a kind of awe. ' Sir,' he exclaimed, ' you would not find another guide in Switzerland to lead you up here.' Nor, indeed, in Bennen's behalf be it spoken, would he have done so if he could have avoided it; but we had fairly got into a net, the meshes of which must be resolutely cut. I had previously entertained the undoubting belief that where a chamois could climb a man could follow; but when I saw the marks of the animal on these all but inaccessible ledges, my belief, though not eradicated, became weaker than it had previously been.

Onward again, slowly winding through the craggy mazes, and closely scanning the cliffs as we ascended. Our easiest work was stiff, but the ' stiff' was an agreeable relaxation from the perilous. By a lateral deviation we reached a point whence we could look into the couloir by which Mr. Tuckett had ascended: here Bennen relieved himself by a sigh and ejaculation: ' Would that we had chosen it! we might pass up yonder rocks blindfold!' But repining was useless; our work was marked out for us and had to be accomplished. After another difficult tug Bennen reached a point whence he could see a large extent

of the rocks above us. There was no serious difficulty within view, and the announcement of this
cheered us mightily. Every vertical yard, however,
was to be won only by strenuous effort. For a long
time the snow cornice hung high above us; we now
approached its level; the last cliff formed a sloping
stair with geologic strata for steps. We sprang up
it, and the magnificent snow-field of the Görner
glacier immediately opened to our view. The
anxiety of the last four hours disappeared like an
unpleasant dream, and with that perfect happiness
which perfect health can alone impart, we consumed
our cold mutton and champagne on the summit of
the old Weissthor.

XIII.

RESCUE FROM A CREVASSE.

MR. HUXLEY and myself had been staying for some
days at Grindelwald, hoping for steady weather, and
looking at times into the wild and noble region
which the Shreckhorn, the Wetterhorn, the Viescher-
hörner, and the Eiger feed with eternal snows.   We
had scanned the buttresses of the Jungfrau with a
view to forcing a passage between the Jungfrau and
the Monk from the Wengern Alp to the Aletsch
glacier.   The weather for a time kept hopes and
fears alternately afloat, but finally it declared against
us; so we moved with the unelastic tread of beaten
soldiers over the Great Sheideck, and up the Vale
of Hasli to the Grimsel.   We crossed the pass
whose planed and polished rocks had long ago attracted the attention of Sir John Leslie, though the
solution which he then offered ignored the ancient
glacier which we now know to have been the planing
tool employed.   On rounding an angle of the Mayenwand, two travellers suddenly appeared in front of
us; they were Mr. (now Sir John) Lubbock and his

guide. He had been waiting at the new hotel erected by M. Seiler at the foot of the Mayenwand, expecting our arrival; and finally, despairing of this, he had resolved to abandon the mountains, and was now bound for Brientz. In fact, the lakes of Switzerland, and the ancient men who once bivouacked along their borders, were to him the principal objects of interest; and we caught him in the act of declaring a preference for the lowlands which we could not by any means share.

We reversed his course, carried him with us down the mountain, and soon made ourselves at home in M. Seiler's hotel. Here we had three days' training on the glacier and the adjacent heights, and on one of the days Lubbock and myself made an attempt upon the Galenstock. By the flank of the mountain, with the Rhone glacier on our right, we reached the heights over the ice cascade and crossed the glacier above the fall. The sky was clear and the air pleasant as we ascended; but in the earth's atmosphere the sun works his swiftest necromancy, the lightness of air rendering it in a peculiar degree capable of change. Clouds suddenly generated came drifting up the valley of the Rhone, covering the glacier and swathing the mountain-tops, but leaving clear for a time the upper *névé* of the Rhone. Grandeur is conceded while beauty is sometimes denied to the Alps. But the higher snow-fields of the great glaciers are altogether beautiful — not throned in repellent grandeur, but endowed with a grace so tender as to suggest the loveliness of woman.

The day was one long succession of surprises wrought by the cloud-filled and wind-rent air. We reached the top, and found there a gloom which might be felt. It was almost thick enough to cut each of us away from the vision of his fellows. But suddenly, in the air above us, the darkness would melt away, and the deep blue heaven would reveal itself spanning the dazzling snows. Beyond the glacier rose the black and craggy summit of the Finsteraarhorn, and other summits and other crags emerged in succession as the battle-clouds rolled away. But the smoke would again whirl in upon us, and we looked once more into infinite haze from the cornice which lists the mountain-ridge. Again the clouds are torn asunder, and again they close. And thus, in upper air, did the sun play a wild accompaniment to the mystic music of the world below

From the Rhone glacier we proceeded down the Rhone valley to Viesch, whence, in the cool twilight, all three of us ascended to the Hotel Jungfrau, on the Æggischhorn. This we made our head-quarters for some days, and here Lubbock and I decided to ascend the Jungfrau. The proprietor of the hotel keeps guides for this excursion, but his charges are so high as to be almost prohibitory. I, however, needed no guide in addition to my faithful Bennen; but simply a porter of sufficient strength and skill to follow where he led. In the village of Laax Bennen found such a porter—a young man named Bielander, who had the reputation of being both courageous and strong. He was the only son of his mother, and she was a widow.

This young man and a second porter we sent on with our provisions to the Grotto of the Faulberg, where we were to spend the night. Between the Æggischhorn and this cave the glacier presents no difficulty which the most ordinary caution cannot overcome, and the thought of danger in connection with it never occurred to us. An hour and a half after the departure of our porters we slowly wended our way to the lake of Märjelin, which we skirted, and were soon upon the ice. The middle of the glacier was almost as smooth as a carriage-road, cut here and there by musical brooks produced by the superficial ablation. To Lubbock the scene opened out with the freshness of a new revelation, as, previously to this year, he had never been among the glaciers of the Alps. To me, though not new, the region had lost no trace of the interest with which I first viewed it. We moved briskly along the frozen incline, until, after a couple of hours' march, we saw a solitary human being standing on the lateral moraine of the glacier, near the point where we were to quit it for the cave of the Faulberg.

At first this man excited no attention. He stood and watched us, but did not come towards us, until finally our curiosity was aroused by observing that he was one of our own two men. The glacier here is always cut by crevasses, which, while they present no real difficulty, require care. We approached our porter, but he never moved; and when we came up to him he looked stupid, and did not speak until he was spoken to. Bennen addressed him in the patois of the place, and he answered in the same patois. His answer must have been more than usually obscure, for Bennen misunderstood the most important part of it. 'My God!' he exclaimed, turning to us, 'Walters is killed!' Walters was the guide at the Æggischhorn, with whom, in the present instance, we had nothing to do. 'No, not Walters,' responded the man; 'it is my comrade that is killed.' Bennen looked at him with a wild bewildered stare. 'How killed?' he exclaimed. 'Lost in a crevasse,' was the reply. We were all so stunned that for some moments we did not quite seize the import of the terrible statement. Bennen at length tossed his arms in the air, exclaiming, 'Jesu Maria! what am I to do?' With the swiftness that some ascribe to dreams, I surrounded the fact with imaginary adjuncts, one of which was

L

that the man had been drawn dead from the crevasse, and was now a corpse in the cave of the Faulberg; for I took it for granted that, had he been still entombed, his comrade would have run or called for our aid. Several times in succession the porter affirmed that the missing man was certainly dead. 'How does he know that he is dead?' Lubbock demanded. 'A man is sometimes rendered insensible by a fall without being killed.' This question was repeated in German, but met with the same dogmatic response. 'Where is the man?' I asked. 'There,' replied the porter, stretching his arm towards the glacier. 'In the crevasse?' A stolid 'Ja!' was the answer. It was with difficulty that I quelled an imprecation. 'Lead the way to the place, you blockhead,' and he led the way.

We were soon beside a wide and jagged cleft which resembled a kind of cave more than an ordinary crevasse. This cleft had been spanned by a snow bridge, now broken, and to the edge of which footsteps could be traced. The glacier at the place was considerably torn, but simple patience was the only thing needed to unravel its complexity. This quality our porter lacked, and, hoping to make shorter work of it, he attempted to cross the bridge. It gave way, and he went down, carrying an immense load of *débris* along with him. We looked into the hole, at one end of which the vision was cut short

by darkness, while immediately under the broken
bridge it was crammed with snow and shattered
icicles. We saw nothing more. We listened with
strained attention, and from the depths of the
glacier issued a low moan. Its repetition assured
us that it was no delusion—the man was still alive.
Bennen from the first had been extremely excited;
and the fact of his having, as a Catholic, saints
and angels to appeal to, augmented his emotion.
When he heard the moaning he became almost
frantic. He attempted to get into the crevasse,
but was obliged to recoil. It was quite plain that a
second life was in danger, for my guide seemed to
have lost all self-control. I placed my hand heavily
upon his shoulder, and admonished him that upon
his coolness depended the life of his friend. 'If
you behave like a man, we shall save him; if like
a woman, he is lost.'

A first-rate rope accompanied the party, but unhappily it was with the man in the crevasse. Coats,
waistcoats, and braces were instantly taken off and
knotted together. I watched Bennen while this
work was going on; his hands trembled with excitement, and his knots were evidently insecure.
The last junction complete, he exclaimed, 'Now let
me down!' 'Not until each of these knots has been
tested; not an inch!'[1] Two of them gave way, and

[1] 'Ach, Herr,' he replied to one of my remonstrances, 'Sein Sie
nicht so hart.'

Lubbock's waistcoat also proved too tender for the strain. The *débris* was about forty feet from the surface of the glacier, but two intermediate prominences afforded a kind of footing. Bennen was dropped down upon one of these; I followed, being let down by Lubbock and the other porter. Bennen then descended the remaining distance, and was followed by me. More could not find room.

The shape and size of the cavity were such as to produce a kind of resonance, which rendered it difficult to fix the precise spot from which the sound issued; but the moaning continued, becoming to all appearance gradually feebler. Fearing to wound the man, the ice-rubbish was cautiously rooted away; it rang curiously as it fell into the adjacent gloom. A layer two or three feet thick was thus removed; and finally, from the frozen mass, and so bloodless as to be almost as white as the surrounding snow, issued a single human hand. The fingers moved. Round it we rooted, cleared the arm, and reached the knapsack, which we cut away. We also regained our rope. The man's head was then laid bare, and my brandy-flask was immediately at his lips. He tried to speak, but his words jumbled themselves to a dull moan. Bennen's feelings got the better of him at intervals; he wrought like a hero, but at times he needed guidance and stern admonition. The arms once free, we passed the

RECOVERY OF OUR PORTER.

rope underneath them, and tried to draw the man out. But the ice-fragments round him had regelated so as to form a solid case. Thrice we essayed to draw him up, thrice we failed; he had literally to be hewn out of the ice, and not until his last foot was extricated were we able to lift him. By pulling him from above, and pushing him from below, the man was at length raised to the surface of the glacier.

For an hour we had been in the crevasse in shirt-sleeves—the porter had been in it for two hours—and the dripping ice had drenched us. Bennen, moreover, had worked with the energy of madness, and now the reaction came. He shook as if he would fall to pieces; but brandy and some dry covering revived him. The rescued man was help-less, unable to stand, unable to utter an articulate sentence. Bennen proposed to carry him down the glacier towards home. Had this been attempted, the man would certainly have died upon the ice. Bennen thought he could carry him for two hours; but the guide underrated his own exhaustion and overrated the vitality of the porter. 'It cannot be thought of,' I said: 'to the cave of Faulberg, where we must tend him as well as we can.' We got him to the side of the glacier, where Bennen took him on his back; in ten minutes he sank under his load. It was now my turn, so I took the man on my back and plodded on with him as far as I was able. Helping each other thus by turns, we reached the mountain grot.

The sun had set, and the crown of the Jungfrau was embedded in amber light. Thinking that the Märjelin See might be reached before darkness, I proposed starting in search of help. Bennen protested against my going alone, and I thought I noticed moisture in Lubbock's eye. Such an occasion brings out a man's feeling if he have any. I gave them both my blessing and made for the glacier. But my anxiety to get quickly clear of the crevasses defeated its own object. Thrice I found myself in difficulty, and the light was visibly departing. The conviction deepened that persistence would be folly, and the most impressive moment of my existence was that on which I stopped at the brink of a profound fissure and looked upon the mountains and the sky. The serenity was perfect—not a cloud, not a breeze, not a sound, while the last hues of sunset spread over the solemn west.

I returned; warm wine was given to our patient, and all our dry clothes were wrapped around him. Hot-water bottles were placed at his feet, and his back was briskly rubbed. He continued to groan a long time; but, finally, both this and the trembling ceased. Bennen watched him solemnly, and at length muttered in anguish, 'Sir, he is dead!' I leaned over the man and found him breathing

gently; I felt his pulse—it was beating tranquilly.
'Not dead, dear old Bennen; he will be able to
crawl home with us in the morning.' The prediction was justified by the event; and two days
afterwards we saw him at Laax, minus a bit of his
ear, with a bruise upon his cheek, and a few scars
upon his hand, but without a broken bone or serious
injury of any kind.

The self-denying conduct of the second porter
made us forget his stupidity—it may have been
stupefaction. As I lay there wet, through the long
hours of that dismal night, I almost registered a
vow never to tread upon a glacier again. But, like
the forces in the physical world, human emotions
vary with the distance from their origin, and a year
afterwards I was again upon the ice.

Towards the close of 1862 Bennen and myself
made 'the tour of Monte Rosa,' halting for a day
or two at the excellent hostelry of Delapierre, in
the magnificent Val du Lys. We scrambled up the
Grauhaupt, a point exceedingly favourable to the
study of the conformation of the Alps. We also
halted at Alagna and Macugnaga. But, notwithstanding their admitted glory, the Italian valleys
did not suit either Bennen or me. We longed
for the more tonic air of the northern slopes, and
were glad to change the valley of Ansasca for

that of Saas. We subsequently, on a perfect day, crossed the Alphubel Joch — a very noble pass, and by no means difficult if the ordinary route be followed. But Bennen and I did not follow that route. We tried to cross the mountains obliquely from the chalets of Täsch, close under the Alphubel, and, as a consequence, encountered on a spur of the mountain a danger to which I will not further refer than to say that Bennen's voice is still present to me as he said, 'Ach, Herr! es thut mir Leid, Sie hier zu sehen.'[1]

[1] Rendered in accordance with the tone and sentiment, this would be, 'Ah! sir, it breaks my heart to see you here.'

XIV.

THE MATTERHORN—SECOND ASSAULT.

FOUR years ago I had not entertained a wish or a thought regarding the climbing of the Matterhorn. Indeed, assailing mountains of any kind was then but an accidental interlude to less exciting occupations upon the glaciers of the Alps. But in 1859 Mr. Vaughan Hawkins had inspected the mountain from Breuil, and in 1860, on the strength of this inspection, he invited me to join him in an attack upon the untrodden peak. Guided by Johann Bennen, and accompanied by an old chamois-hunter named Carrel, we tried the mountain, but had to halt midway among its precipices. We returned to Breuil with the belief that, if sufficient time could be secured, the summit—at least, *one* summit—might be won. Had I felt that we had done our best on this occasion, I should have relinquished all further thought of the mountain; but, unhappily, I felt the reverse, and thus a little cloud of dissatisfaction hung round the memory of the attempt. In 1861 I once more looked at the Matterhorn, but, as shown in Chapter X., was forbidden to set foot upon it. Finally, in 1862, the desire to finish what I felt to be a piece of work only half completed beset me so strongly that I resolved to make a last attack upon the unconquered hill.

The resolution, as a whole, may have been a rash one, but there was no rashness displayed in the carrying out of its details. I did not ignore the law of gravity, but felt, on the contrary, that the strongest aspirations towards the summit of the Matterhorn would not prevent precipitation to its base through a false step or a failing grasp. The general plan proposed was this: Two first-rate guides were to be engaged, and, to leave their arms free, they were to be accompanied by two strong and expert porters. The party was thus to consist of five in all. During the ascent it was proposed that three of those men should always be, not only out of danger, but attached firmly to the rocks; and while they were thus secure, it was thought that the remaining two might take liberties, and commit themselves to ventures which would otherwise be inexcusable or impossible. With a view to this, I had a rope specially manufactured in London, and guaranteed by its maker to bear a far greater strain than was ever likely to be thrown upon it. A light ladder was also constructed, the two sides of which might be carried like huge alpenstocks, while its steps, which could be inserted at any moment, were strapped upon a porter's back. Long iron nails and a hammer were also among our appliances. Actual experience considerably modified these arrangements, and compelled us in almost all cases to resort to methods as much open to a savage as to people acquainted with the mechanical arts.

Throughout the latter half of July rumours from the Matterhorn were rife in the Bernese Oberland, and I felt an extreme dislike to add to the gossip. Wishing, moreover, that others who desired it might have a fair trial, I lingered for nearly three weeks among the Bernese and Valasian Alps. This time, however, was not wasted. It was employed in burning up the effete matters which nine months' work in London had lodged in my muscles—in rescuing the blood from that fatty degeneration which a sedentary life is calculated to induce. I chose instead of the air of a laboratory that of the Wetterhorn, the Galenstock, and the mountains which surround the Great Aletsch glacier. Each succeeding day added to my strength.

There is assuredly morality in the oxygen of the mountains, as there is immorality in the miasma of a marsh, and a higher power than mere brute force lies latent in Alpine mutton. We are recognising more and more the influence of physical elements in the conduct of life, for when the blood

flows in a purer current the heart is capable of
a higher glow. Spirit and matter are interfused;
the Alps improve us *totally*, and we return from
their precipices wiser as well as stronger men.

It is usual for the proprietor of the hotel on the
Æggischhorn to retain a guide for excursions in the
neighbourhood; and last year he happened to have
in his employment one Walters, a man of superior
strength and energy. He was the house companion
of Bennen, who was loud in his praise. Thinking
it would strengthen Bennen, hand and heart, to
have so tried a man beside him, I engaged Walters,
and we all three set off with cheerful spirits to
Zermatt. Thence we proceeded over the Matter-
joch; and as we descended to Breuil we looked
long at the dangerous eminences to our right,
among which we were to trust ourselves in a day or
two. There was nothing jubilant in our thoughts
or conversation; the character of the work before
us quelled presumption. We felt nothing that
could be called confidence as to the issue of the
enterprise, but we also felt the inner compactness
and determination of men who, though they know
their work to be difficult, feel no disposition to
shrink from it. The Matterhorn, in fact, was our
temple, and we aproached it with feelings not unworthy of so great a shrine. Arrived at Breuil, we
found that a gentleman, whose long perseverance

merited victory (and who has since gained it),[1] was
then upon the mountain. The succeeding day was
spent in scanning the crags and in making prepara-
tions. At night Mr. Whymper returned from the
Matterhorn, having left his tent upon the rocks. In
the frankest spirit, he placed it at my disposal, and
thus relieved me from the necessity of carrying up
my own. At Breuil I engaged two porters, both
named Carrel, the youngest of whom was the son
of the Carrel who accompanied Mr. Hawkins and
me in 1860, while the other was old Carrel's nephew.
He had served as a *bersaglier* in three campaigns,
and had fought at the battle of Solferino; his
previous habits of life rendered him an extremely
handy and useful companion, and his climbing
powers proved also very superior.

About noon on an August day we disentangled
ourselves from the hotel, first slowly sauntering
along a small green valley, but soon meeting the
bluffs, which indicated our approach to uplifted
land. The bright grass was quickly left behind, and
soon afterwards we were toiling laboriously upward
among the rocks. The Val Tournanche is bounded
on the right by a chain of mountains, the higher
end of which abutted, in former ages, against
Matterhorn. But now a gap is cut out between
both, and a saddle stretches from the one to the

[1] Mr. Whymper.

other. From this saddle a kind of couloir runs downwards, widening out gradually and blending with the gentler slopes below. We held on to the rocks to the left of this couloir, until we reached the base of a precipice which fell sheer from the summits above. Water trickled from the upper ledges, and the descent of a stone at intervals admonished us that gravity had here more serious missiles at command than the drippings of the liquefied snow. So we moved with prudent speed along the base of the precipice, crossing at one place the ice-gulley where Mr. Whymper nearly lost his life. Immediately afterwards we found ourselves upon the saddle which stretches with the curvature of a chain to the base of the true Matterhorn. The opening out of the western mountains from this point of view is grand and impressive, and with our eyes and hearts full of the scene we moved along the saddle, and soon came to rest upon the first steep crags of the real Monarch of the Alps.

Here we paused, unlocked our scrip, and had some bread and wine. Again and again we looked to the cliffs above us, ignorant of the treatment that we were to receive at their hands. We had gathered up our traps, and bent to the work before us, when suddenly an explosion occurred overhead. We looked aloft and saw in mid-air a solid shot from the Matterhorn, describing its proper parabola, and

finally splitting into fragments as it smote one of
the rocky towers in front of us. Down the scattered
fragments came like a kind of spray, slightly wide
of us, but still near enough to compel a sharp look-
out. Two or three such explosions occurred, but we
chose the back-fin of the mountain for our track,
and from this the falling stones were speedily
deflected right or left. Before the set of sun we
reached our place of bivouac. A roomy tent was
already there, and we had brought with us an
additional light one, intended to afford accom-
modation to me. It was pitched in the shadow of
a great rock, which seemed to offer a safe barrier
against the cannonade from the heights. Carrel,
the soldier, built a platform, on which he placed the
tent, for the mountain itself furnished no level
space of sufficient area.

Meanwhile, fog, that enemy of the climber, came
creeping up the valleys, while dense flounces of
cloud gathered round the hills. As night drew
near, the fog thickened through a series of inter-
mittences which a mountain-land alone can show.
Sudden uprushings of air would often carry the
clouds aloft in vertical currents, while horizontal
gusts swept them wildly to and fro. Different
currents impinging upon each other sometimes
formed whirling cyclones of cloud. The air was
tortured in its search of equilibrium. Sometimes

all sight of the lower world was cut away—then again the fog would melt and show us the sunny pastures of Breuil smiling far beneath. Sudden peals upon the heights, succeeded by the sound of tumbling rocks, announced, from time to time, the disintegration of the Matterhorn. We were quite swathed in fog when we retired to rest, and had scarcely a hope that the morrow's sun would be able to dispel the gloom. Throughout the night the rocks roared intermittently, as they swept down the adjacent couloir. I opened my eyes at midnight, and, through a minute hole in the canvas of my tent, saw a star. I rose and found the heavens swept clear of clouds, while above me the solemn battlements of the Matterhorn were projected against the blackened sky.

At 2 A.M. we were astir. Carrel made the fire, boiled the water, and prepared our coffee. It was 4 A.M. before we had fairly started. We adhered as long as possible to the hacked and weather-worn spine of the mountain, until at length its disintegration became too vast. The alternations of sun and frost have made wondrous havoc on the southern face of the Matterhorn; but they have left brown-red masses of the most imposing magnitude behind—pillars, and towers, and splintered obelisks, grand in their hoariness—savage, but still softened by the colouring of age. The mountain is a gigantic ruin;

but its firmer masonry will doubtless bear the shocks of another æon. We were compelled to quit the ridge, which now swept round and fronted us like a wall. The weather had cleft the rock clean away, leaving smooth sections, with here and there a ledge barely competent to give a man footing. It was manifest that for some time our fight must be severe. We examined the precipice, and exchanged opinions. Bennen swerved to the right and to the left to render his inspection complete. There was no choice; over this wall we must go, or give up the attempt. We reached its base, roped ourselves together, and were soon upon the face of the precipice. Walters was first, and Bennen second, both exchanging pushes and pulls. Walters, holding on to the narrow ledges above, scraped his ironshod boots against the cliff, thus lifting himself in part by friction. Bennen was close behind, aiding him with an arm, a knee, or a shoulder. Once upon a ledge, he was able to give Bennen a hand. Thus we advanced, straining, bending, and clinging to the rocks with a grasp like that of desperation, but with heads perfectly cool. We perched upon the ledges in succession—each in the first place making his leader secure, and accepting his help afterwards. A last strong effort threw the body of Walters across the top of the wall; and, he being safe, our success thus far was ensured.

M

This ascent landed us once more upon the ridge, with safe footing on the ledged strata of the disintegrated gneiss. Pushing upward, we approached the conical summit seen from Breuil—the peak, however, being the end of a nearly horizontal ridge foreshortened from below. But before us, and assuredly as we thought within our grasp, was the highest point of the renowned Matterhorn. 'Well,' I remarked to Bennen, 'we shall at all events win the lower summit.' 'That will not satisfy us,' was his reply. I knew he would answer thus, for a laugh of elation, which had something of scorn in it, had burst from the party when the true summit came in view. We felt perfectly certain of success; not one amongst us harboured a thought of failure. 'In an hour,' cried Bennen, 'the people at Zermatt shall see our flag planted yonder.' Up we went in this spirit, with a forestalled triumph making our ascent a jubilee. We reached the first summit, and planted a flag upon it. Walters, however, who was an exceedingly strong and competent guide, but without the genius which is fired by difficulty, had previously remarked, with reference to the last precipice of the mountain, 'We may still find difficulty there.' The same thought had probably brooded in other minds; still it angered me slightly to hear misgiving obtain audible expression.

From the point on which we planted our first

flagstaff a hacked and extremely acute ridge ran, and abutted against the final precipice. Along this we moved cautiously, while the face of the precipice came clearer and clearer into view. The ridge on which we stood ran right against it; it was the only means of approach, while ghastly abysses fell on either side. We sat down, and inspected the place: no glass was needed, it was so near. Three out of the four men muttered almost simultaneously, 'It is impossible.' Bennen was the only man of the four who did not utter the word. A jagged stretch of the ridge still separated us from the precipice. I pointed to a spot at some distance from the place where we sat, and asked the three doubters whether that point might not be reached without much danger. 'We think so,' was the reply. 'Then let us go there.' We reached the place, and sat down there. The men again muttered despairingly, and at length they said distinctly, 'We must give it up.' I by no means wished to put on pressure, but directing their attention to a point at the base of the precipice, I asked them whether they could not reach that point without much risk. The reply was, 'Yes.' 'Then,' I said, 'let us go there.' We moved cautiously along, and reached the point aimed at. The ridge was here split by a deep cleft which separated it from the final precipice. So savage a spot I had never seen, and I sat down upon it with the sickness of disappointed hope. The summit was within almost a stone's throw of us, and the thought of retreat was bitter in the extreme. Bennen excitedly pointed out a track which he thought practicable. He spoke of danger, of difficulty, never of impossibility; but this was the ground taken by the other three men.

As on other occasions, my guide sought to fix on me the responsibility of return, but with the usual result. 'Where you go I will follow, be it up or down.' It took him half an hour to make up his mind. But he was finally forced to accept defeat. What could he do? The other men had yielded utterly, and our occupation was clearly gone. Hacking a length of six feet from one of the sides of our ladder, we planted it on the spot where we stopped. It was firmly fixed, and, protected as it is from lightning by the adjacent peak, it will probably stand there when those who planted it think no more of the Matterhorn.

How this wondrous mountain has been formed will be the subject of subsequent enquiry. It is not a spurt of molten matter ejected from the nucleus of the earth; from base to summit there is no truly igneous rock. It has no doubt been upraised by subterranean forces, but that it has been lifted as an isolated mass is not conceivable. It must have formed part of a mighty boss or swelling, from which the mountain was subsequently sculptured. These subjects, however, cannot be well discussed here. To get down the precipice we had scaled in the morning, we had to fix the remaining length of our ladder at the top, to tie our rope firmly on to it, and allow it to hang down the cliff. We slid down it in succession, and there it still dangles, for we could not detach it. A tempest of hail was here hurled against us; as if the Matterhorn, not content with shutting its door in our faces, meant to add an equivalent to the process of kicking us downstairs. The ice-pellets certainly hit us as bitterly as if they had been thrown in spite, and in the midst of this malicious cannonade we struck our tents and returned to Breuil.

[Three years subsequently, Carrel the *bersaglier*, and some other Val Tournanche men, reached my rope, found it bleached to whiteness, but still strong enough to bear the united weights of three men. By it they were enabled to scale the precipice, spend the night at a considerable elevation, and, through the scrutiny rendered possible by an early start, to find a way to the summit of the Matterhorn. They reached the top a day or two after the memorable first ascent from Zermatt.]

ADDITIONAL NOTE.

I WAS by no means satisfied with the cursory inspection
of the Matterhorn in 1861, and during the ensuing winter
I wrote to Bennen expressing this feeling and proposing a
really earnest attack upon the mountain. He resolutely
closed with my proposal and hence the foregoing excursion.
To Mr. Good, of King William Street, I was recommended
by the late Mr. Appold, the celebrated mechanician ; and
Mr. Good then manufactured for me the excellent rope
employed on the Matterhorn in 1862. Through the kindness of Carrel two or three feet of this rope, blanched by
three years' exposure, are now in my possession. I should
have been at Breuil a fortnight earlier in 1862, had I not
wished to leave Mr. Whymper in undisturbed possession of
the mountain, and not until I had heard that his attack
had been ended by a serious fall did I cross the Theodule.
I was rejoiced to find him active, though his wounds were
by no means healed. It certainly would have enhanced the
pleasure of my excursion if Mr. Whymper could have
accompanied me. I admired his courage and devotion ; he
had manifestly set his heart upon the Matterhorn, and it
was my earnest desire that he should not be disappointed.
I consulted with Bennen, who had heard many accounts—
probably exaggerated ones—of Mr. Whymper's rashness.
He shook his head, but finally agreed that Mr. Whymper
should be invited, ' provided he proved reasonable.' I
thereupon asked Mr. Whymper to join us: his reply was,
' If I go up the Matterhorn I must lead the way.' Con-

sidering my own experience at the time as compared with his; considering, still more, the renown and power of my guide, I thought the response the reverse of 'reasonable,' and so went on my way alone.

In the praise of Carrel as a climber, in which Mr. Whymper indulges in his recent magnificent volume, I heartily join. It would be difficult to find two better rock-men than his colleague Joseph Maquignaz and himself. But I am concerned to find Carrel originating, and still more concerned to find Mr. Whymper giving currency to the fiction, that when Bennen halted in 1862 at the base of the final precipice of the Matterhorn, he, Carrel, stood forth and pointed out a way to the top. Of the guides and porters Bennen was the only man who entertained a thought of going on ; he was throughout the natural commander of the party, and both Walters and Carrel shrunk from the dangers of the last ascent. But with the experience derived from his association with Bennen, Carrel three years afterwards returned to the charge. He had my rope to aid him, and by it he and his colleagues readily reached an elevation which gave them ample time for the thorough inspection of the last bit of the mountain. But even thus aided he accomplished a task which none but a first-rate climber could have accomplished, and it is doubly unworthy of a man capable of so gallant an achievement to deal untruly with the memory of a heroic predecessor, without whose initiative he might never have set his foot upon the Matterhorn.

Regarding other inaccuracies, and touching Mr. Whymper's general tone towards myself, I do not feel called upon to make any observations.

July, 1871.

XV.

FROM STEIN TO THE GRIMSEL.

On the 19th of July 1863 Mr. Philip Lutley Sclater and I reached Reichenbach, and on the following day we sauntered up the valley of Hasli, and over the Kirchet to Imhof, where we turned to the left into Gadmenthal.   Our destination was Stein, which we reached by a grass-grown road through fine scenery.   The goatherds were milking when we arrived.   At the heels of one quadruped, supported by the ordinary uni-legged stool of the Senner, bent a particularly wild and dirty-looking individual, who, our guide informed us, was the proprietor of the inn.   'He is but a rough Bauer,' said Jaun, 'but he has engaged a pretty maiden to keep house for him.'   While he thus spoke a light-footed creature glided from the door towards us, and bade us welcome.   She led us upstairs, provided us with baths, took our orders for dinner, helped us by her suggestions, and answered all our questions   with the utmost propriety and grace.   She had been two years in England, and

spoke English with a particularly winning accent.
How she came to be associated with the unkempt
individual outside was a puzzle to both of us.   It is
Emerson, I think, who remarks on the benefit which
a beautiful face, without trouble to itself, confers
upon him who looks at it.   And, though downright
beauty could hardly be claimed for our young
hostess, she was handsome enough and graceful
enough to brighten a tired traveller's thoughts, and
to raise by her presence the modest comforts she
dispensed to the level of luxuries.[1]

It rained all night, and at 3.30 A.M., when we were
called, it still fell heavily.   At 5, however, the
clouds began to break, and half an hour afterwards
the heavens were swept quite clear of them.   At 6
we bade our pretty blossom of the Alps good-bye.
She had previously brought her gentle influence to
bear upon her master to moderate the extortion of
some of his charges.   We were soon upon the Stein
glacier, and after some time reached a col from

---

[1] Thackeray, in his 'Peg of Limavady,' is perhaps more to the
point than Emerson:

> 'Presently a maid
>   Enters with the liquor—
> Half a pint of ale
>   Frothing in a beaker;
> As she came she smiled,
>   And the smile bewitching,
> On my word and honour,
>   Lighted all the kitchen.'

which we looked down upon the lower portion of the nobler and more instructive Trift glacier. Brown bands were drawn across the ice-stream, forming graceful loops with their convexities turned downwards. The higher portions of the glacier were not in view; still those bands rendered the inference secure that an ice-fall existed higher up, at the base of which the bands had originated. We shot down a shingly couloir to the Trift, and looking up the glacier the anticipated cascade came into view. At its bottom the ice, by pressure, underwent that notable change, analogous to slaty cleavage, which caused the glacier to weather and gather dirt in parallel grooves, thus marking upon its surface the direction of its interior lamination.

The ice-cascade being itself impracticable, we scaled the rocks to the left of it, and were soon in presence of the far-stretching snow-fields from which the lower glacier derives nutriment. With a view to hidden crevasses, we here roped ourselves together. The sun was strong, its direct and reflected blaze combining against us. The scorching warmth experienced at times by cheeks, lips, and neck, indicated that in my case mischief was brewing; but the eyes being well protected by dark spectacles, I was comparatively indifferent to the prospective disfigurement of my face. Mr. Sclater was sheltered by a veil, a mode of defence which the habit of going into places requiring the unimpeded eyesight has caused me to neglect.

There would seem to be some specific quality in the sun's rays which produces the irritation of the skin experienced in the Alps. The solar heat may be compared, in point of *quantity*, with that radiated from a furnace; and the heat encountered by the mountaineer on Alpine snows is certainly less intense than that endured by workmen in many of our technical operations. But the terrestrial heat appears to lack the *quality* which gives the solar rays their power. The sun is incomparably richer in what are called chemical rays than are our fires, and to such rays the irritation may be due. The keen air of the heights may also have much to do with it. As a remedy for sunburn I have tried glycerine, and found it a failure. The ordinary lip-salve of the druggists' shops is also worse than useless; but pure cold-cream, for a supply of which I have often had occasion to thank a friend, is an excellent ameliorative.

After considerable labour we reached the ridge— a very glorious one as regards the view—which forms the common boundary of the Rhone and Trift glaciers.[1] Before us and behind us for many a mile fell the dazzling *névés*, down to the points where the

---

[1] Seven years previously Mr. Huxley and myself had attempted to reach this col from the other side.

grey ice emerging from its white coverlet declared
the junction of snow-field and glacier. We had
plodded on for hours soddened by the solar heat and
parched with thirst. There was

> Water, water everywhere,
> But not a drop to drink ;

for, when placed in the mouth, the liquefaction of
the ice was so slow, and the loss of heat from the
surrounding tissues so painful, that sucking it was
worse than total abstinence. In the midst of this
solid water you might die of thirst. At some distance below the col, on the Rhone side, the musical
trickle of water made itself audible, and to the
rocks from which it fell we repaired, and refreshed
ourselves. The day was far spent, the region was
wild and lonely, when, beset by that feeling which
has often caused me to wander singly in the Alps, I
broke away from my companions, and went rapidly
down the glacier. Our guide had previously informed me that before reaching the cascade of the
Rhone the ice was to be forsaken, and the Grimsel,
our destination, reached by skirting the base of the
peak called Nägelis Grätli. After descending the
ice for some time I struck the bounding rocks, and,
climbing the mountain obliquely, found myself among
the crags which lie between the Grimsel pass and
the Rhone glacier. It was an exceedingly desolate

place, and I soon had reason to doubt the wisdom of being there alone. Still difficulty rouses powers of which we should otherwise remain unconscious. The heat of the day had rendered me weary, but among these rocks the weariness vanished, and I became clear in mind and fresh in body through the knowledge that after nightfall escape from this wilderness would be impossible.

I reached the watershed of the region, where I accepted the guidance of a tiny stream. It received in its course various lateral tributaries, and at one place expanded into a small blue lake bounded by banks of snow. I kept along its side afterwards until, arching over a brow of granite, it discharged itself down precipitous and glaciated rocks. Here I learned that the stream was the feeder of the Grimsel lake. I halted on the brow for some time. The hospice was in sight, but the precipices between it and me seemed desperately forbidding. Nothing is more trying to the climber than those cliffs which have been polished by the ancient glaciers. Even at moderate inclinations, as may be learned from an experiment on the Höllenplatte, or some other of the polished rocks in Haslithal, they are not easy. I need hardly say that the inclination of the rocks flanking the Grimsel is the reverse of moderate. It is dangerously steep.

How to get down these smooth and precipitous

tablets was now a problem of the utmost interest to me ; for the day was too far gone, and I was too ignorant of the locality, to permit of time being spent in the search of an easier place of descent. Right or left of me I saw none. The continuity of the cliffs below me was occasionally broken by cracks and narrow ledges, with scanty grass-tufts sprouting from them here and there. The problem was to get down from crack to crack and from ledge to ledge. A salutary anger warms the mind when thus challenged, and, aided by this warmth, close scrutiny will dissolve difficulties which timidity might render insuperable. Bit by bit I found myself getting lower, closely examining at every pause the rocks below me. The grass-tufts helped me for a time, but at length a slab was reached where no friendly grass could grow. This slab was succeeded by others equally forbidding. I looked upwards, thinking of retreat, but the failing day urged me on. From the middle of the smooth surface jutted a narrow ledge. Grasping the top of the rock, I let myself down as far as my stretched arms would permit, and then let go my hold. I came upon the ledge with an energy that alarmed me. A downward-pointing crack with a streak of grass in it was next attained ; it terminated in a small, steep gulley, the portion of which within view was crossed by three transverse ledges, and I judged that by friction the motion down the groove could be so regulated as to enable me to come to rest at each successive ledge. But the rush was unexpectedly rapid, and I shot over the first ledge. Having some power in reserve, I tried to clamp myself against the rock, but the second ledge was crossed like the first. The wish to spare clothes or avoid abrasions of the skin here vanished, and for dear life I grappled with the rock. Braces gave way, clothes were rent, wrists and hands were skinned and bruised, while hips and knees suffered variously. The motion however ended. I was greatly heated, but immensely relieved otherwise. A little lower down I discovered a singular cave in the mountain-side, with water dripping from its roof into a well of crystal clearness. The ice-cold liquid soon restored me to a normal temperature. I felt quite fresh on entering the Grimsel inn, but a curious physiological effect manifested itself when I had occasion to speak. The power of the brain over the lips was so lowered that I could hardly make myself understood.

XVI.

THE OBERAARJOCH.

ADVENTURE AT THE ÆGGISCHHORN.

MY guide Bennen reached the Grimsel the following morning. Uncertain of my own movements, I had permitted him this year to make a new engagement, which he was now on his way to fulfil. As a mountaineer, Bennen had no superior, and he added to his strength, courage, and skill, the qualities of a natural gentleman. He was now ready to bear us company over the Oberaarjoch to the Æggischhorn. On the morning of the 22nd we bade the cheerless Grimsel inn good-bye, reached the Unteraar glacier, crossed its load of uncomfortable moraine shingle, and clambered up the slopes at the other side. Nestled aloft in a higher valley was the Oberaar glacier, along the unruffled surface of which our route lay.

The morning threatened, and fitful gleams of sunlight wandered over the ice. The Joch was swathed in mist, which now and then gave way, permitting a wild radiance to shoot over the top. On the windy summit we took a mouthful of food and roped ourselves together. Here, as in a hundred other places, I sought in the fog for the vesicles of De Saussure, but failed to find them. Bennen, as long as we were on the Berne side of the col, permitted Jaun to take the lead; but now we looked into Wallis, or rather into the fog which filled it, and he, as Wallis guide, came to the front. I knew the Viesch glacier well, but how Bennen meant to unravel its difficulties without landmarks I knew not. I asked him whether, if the fog continued, he could make his way down the glacier. There was a pleasant *timbre* in Bennen's voice, a light and depth in his smile, due to the blending together of conscious strength and warm affection. With this smile he turned round and said, 'Herr, ich bin hier zu Hause. Der Viescher Gletscher ist meine Heimath.'

Downwards we went, striking the rocks of the Rothhorn so as to avoid the riven ice. Suddenly we passed from dense fog into clear air : we had crossed ' the cloud-plane,' and found a transparent atmosphere between it and the glacier. The dense covering above us was sometimes torn asunder by the wind, which whirled the detached cloud-tufts round the peaks. Contending air-currents were thus revealed, and thunder, which is the common associate, if not the product, of such contention, began to rattle among the crags. At first the snow upon the glacier was sufficiently heavy to bridge the

N

crevasses, thus permitting of rapid motion ; but by degrees the fissures opened, and at length drove us to the rocks.   These in their turn became impracticable.   Dropping down a waterfall well known to the climbers of this region, we came again upon the ice, which was here cut by complex chasms.   These we unravelled as long as necessary, and finally escaped from them to the mountain-side.   The first big drops of a thunder-shower were already falling when we reached an overhanging crag which gave us shelter.   We quitted it too soon, beguiled by a treacherous gleam of blue, and were thoroughly drenched before we reached the Æggischhorn.

This was my last excursion with Bennen.   In the month of February of the following year he was killed by an avalanche on the Haut de Cry, a mountain near Sion.[1]

Having work to execute, I remained at the Æggischhorn for nearly a month in 1863.   My favourite place for rest and writing was a point on the mountain-side about an hour westwards from the hotel, where the mighty group of the Mischabel, the Matterhorn, and the Weisshorn were in full view. One day I remained in this position longer than

---

[1] Bennen's death is described in Chapter XVIII. A liberal collection was made in England for his mother and sisters ; and Mr. Hawkins, Mr. Tuckett, and myself, had a small monument erected to his memory in Ernan churchyard.   The supervision of the work was entrusted to a clerical friend of Bennen's, who made but a poor use of his trust.

usual, held by the fascination of the setting sun.
The mountains had stood out nobly clear during the
entire day, but towards evening, upon the Dom, a
singular cloud settled, which was finally drawn into
a long streamer by the wind. Nothing can be finer
than the effect of the red light of sunset on those
streamers of cloud. Incessantly dissipated, but ever
renewed, they glow with the intensity of flames. By-
and-by the banner broke, resembling in its action
that of a liquid cylinder when unduly stretched,
forming a series of crimson cloud-balls which were
united by slender filaments of fire. I waited for this
glory to fade into a deadly pallor before I thought
of returning to the hotel.

On arriving there I found discussed with eager
interest the fate of two ladies and a gentleman, who
had quitted the hotel in the morning without a guide,
and who were now, it was said, lost on the mountain.
' I recommended them,' said Herr Wellig, the landlord, ' to take a guide, but they would not heed me.'
I asked him what force he had at hand. Three
active young fellows came immediately forward.
Two of them I sent across the mountain by the
usual route to the Märjelin See, and the third I
took with myself along the watercourse of the
Æggischhorn. After some walking we dipped into
a little dell, where the glucking of cowbells announced the existence of chalets. The party had

been seen passing there in the morning, but not
returning. The embankment of the watercourse
fell at some places vertically for twenty or thirty
feet. Here I thought an awkward slip might have
occurred, and, to meet the possibility of having to
carry a wounded man, I took an additional lithe
young fellow from the chalet.

We shouted as we went along, but the echoes were
our only response. Our pace was rapid, and in the
dubious light false steps were frequent. We all at
intervals mistook the grey water for the grey and
narrow track beside it, and stepped into the stream.
We proposed ascending to the chalets of Märjelin,
but previous to quitting the watercourse we halted,
and, directing our voices down hill, shouted a last
shout. And faintly up the mountain came a sound
which could not be an echo. We all heard it, though
it could hardly be detached from the murmur of the
adjacent stream. We went rapidly down the Alp,
and after a little time shouted again. More audible
than before, but still very faint, came the answer
from below. We continued at a headlong pace, and
soon assured ourselves that the sound was not only
that of a human voice, but of an English voice.
Thus stimulated, we swerved to the left, and, regardless of a wetting, dashed through the torrent
which tumbles from the Märjelin See. Close to the
Viesch glacier we found the objects of our search—

the two ladies, tired out, seated upon the threshold of a forsaken chalet, and the gentleman seated on a rock beside them.

He was both an experienced climber and a brave man, but he had started with a sprained ankle, and every visitor knows how bewildering the spurs of the Æggischhorn are, even to those whose tendons are sound. Thus weakened, he was overtaken by the night, lost his way, and, in his efforts to extricate himself, had experienced one or two serious tumbles. Finally, giving up the attempt, he had resigned himself to spending the night where we found him. The ladies were quite tired out, and to reach the Æggischhorn that night was out of the question. I tried the chalet door and found it locked, but an ice-axe soon hewed the bolt away, and forced an entrance. There was some pinewood within, and some old hay, which, under the circumstances, formed a delicious couch for the ladies. In a few minutes a fire was blazing and crackling in the chimney corner. Having thus secured them, I returned to the chalets first passed, sent them bread, butter, cheese, and milk, and had the lively gratification of seeing them return safe and sound to the hotel next morning.

XVII.

ASCENT OF THE JUNGFRAU.

I HAD spent nearly a fortnight at the Æggischhorn in 1863, employing alternate days in wandering and musing over the green Alps, and in more vigorous action upon the Aletsch glacier. Day after day a blue sky spanned the earth, and night after night the stars glanced down from an unclouded heaven. There is no nobler mountain group in Switzerland than that seen on a fine day from the middle of the Aletsch glacier looking southwards; while to the north, and more close at hand, rise the Jungfrau and other summits familiar to every tourist who has crossed the Wengern Alp. The love of being alone amid those scenes caused me, on the 3rd of August, to withdraw from all society, and ascend the glacier, which for nearly two hours was almost as even as a highway, no local danger calling away the attention from the near and distant mountains. The ice yielded to the sun, rills were formed, which united to rivulets, and these again coalesced to rapid brooks, which ran with a pleasant music through deep channels cut in the ice. Sooner or later these brooks were crossed by cracks; into these cracks the water fell, scooping gradually out for itself a vertical shaft, the resonance of which raised the sound of the falling water to the dignity of thunder. These shafts constitute the so-called moulins of the glacier, examples of which are shown upon the Mer de Glace to every tourist who visits the Jardin from Chamouni. The moulins can only form where the glacier is not much riven, as here alone the rivulets can acquire the requisite volume to produce a moulin.

After two hours' ascent, the ice began to wear a more hostile aspect, and long stripes of last year's snow drawn over the sullied surface marked the lines of crevasses now partially filled and bridged over. For a time this snow was consolidated, and I crossed numbers of the chasms, sounding in each case before trusting myself to its tenacity. But as I ascended, the width and depth of the fissures increased, and the fragility of the snow-bridges became more conspicuous. The crevasses yawned here and there with threatening gloom, while along their fringes the crystallising power of water played the most fantastic freaks. Long lines of icicles dipped into the darkness, and at some places the liquefied snow had refrozen into clusters of plates, ribbed and serrated like the leaves of ferns.

The cases in which the snow covering of the crevasses, when tested by the axe, yielded, became
gradually more numerous, demanding commensurate
caution. It is impossible to feel otherwise than
earnest in such scenes as this, with the noblest and
most beautiful objects in nature around one, with
the sense of danger raising the feelings at times to
the level of awe.

My way upwards became more and more difficult,
and circuit after circuit had to be made to round
the gaping fissures. There is a passive cruelty in
the aspect of these chasms sufficient to make the
blood run cold. Among them it is not good for
man to be alone, so I halted in the midst of them
and swerved back towards the Faulberg. But instead
of it I struck the lateral tributary of the Aletsch,
which runs up to the Grünhorn Lücke. In this
passage I was more than once entangled in a mesh
of fissures ; but it is marvellous what steady, cool
scrutiny can accomplish upon the ice, and how often
difficulties of apparently the gravest kind may be
reduced to a simple form by skilful examination.
I tried to get along the rocks to the Faulberg, but
after investing half an hour in the attempt I thought
it prudent to retreat. I finally reached the Faulberg by the glacier, and with great comfort consumed
my bread and cheese and emptied my goblet in the
shadow of its caves. On this day it was my desire

to get near the buttresses of the Jungfrau, and to see what prospect of success a lonely climber would have in an attempt upon the mountain. Such an attempt might doubtless be made, but at a risk which no sane man would willingly incur.

On August 6, however, I had the pleasure of joining Dr. Hornby and Mr. Philpotts, who, with Christian Almer and Christian Lauener for their guides, wished to ascend the Jungfrau. We quitted the Æggischhorn at 2.15 P.M., and in less than four hours reached the grottoes of the Faulberg. A pine fire was soon blazing, a pan of water soon bubbling sociably over the flame, and the evening meal was quickly prepared and disposed of. For a time the air behind the Jungfrau and Monk was exceedingly dark and threatening; rain was streaming down upon Lauterbrunnen, and the skirt of the storm wrapped the summits of the Jungfrau and the Monk. Southward, however, the sky was clear, and there were such general evidences of hope that we were not much disheartened by the local burst of ill-temper displayed by the atmosphere to the north of us. Like a gust of passion the clouds cleared away, and before we went to rest all was sensibly clear. Still the air was not transparent, and for a time the stars twinkled through it with a feeble ray. There was no visible turbidity, but a something which cut off half the stellar brilliancy.

The starlight, however, became gradually stronger, not on account of the augmenting darkness, but because the air became clarified as the night advanced.

Two of our party occupied the upper cave, and the guides took possession of the kitchen, while a third lay in the little grot below. Hips and ribs felt throughout the night the pressure of the subjacent rock. A single blanket, moreover, though sufficient to keep out the pain of cold, was insufficient to induce the comfort of warmth; so I lay awake in a neutral condition, neither happy nor unhappy, watching the stars without emotion as they appeared in succession above the mountain-heads.

At half-past 12 a rumbling in the kitchen showed the guides to be alert, and soon afterwards Christian Almer announced that tea was prepared. We rose, consumed a crust and basin each, and at 1.15 A.M., being perfectly harnessed, we dropped down upon the glacier. The crescent moon was in the sky, but for a long time we had to walk in the shadow of the mountains, and therefore required illumination. The bottoms were knocked out of two empty bottles, and each of these, inverted, formed a kind of lantern which protected from the wind a candle stuck in the neck. Almer went first, holding his lantern in his left hand and his axe in the right, moving cautiously along the snow which, as the residue of the spring avalanches, fringed the glacier.

At times, for no apparent reason, the leader paused and struck his ice-axe into the snow. Looking right or left, a chasm was always discovered in these cases, and the cautious guide sounded the snow, lest the fissure should have prolonged itself underneath so as to cross our track. A tributary glacier joined the Aletsch from our right—a long corridor filled with ice, and covered by the purest snow. Down this valley the moonlight streamed, silvering the surface upon which it fell.

Here we cast our lamps away, and roped ourselves together. To our left a second long ice-corridor stretched up to the Lötsch saddle, which hung like a chain between the opposing mountains. In fact, at this point four noble ice-streams form a junction, and flow afterwards in the common channel of the Great Aletsch glacier. Perfect stillness might have been expected to reign upon the ice, but even at that early hour the gurgle of subglacial water made itself heard, and we had to be cautious in some places lest a too thin crust might let us in. We went straight up the glacier, towards the col which links the Monk and Jungfrau together. The surface was hard, and we went rapidly and silently over the snow. There is an earnestness of feeling on such occasions which subdues the desire for conversation. The communion we held was with the solemn mountains and their background of dark blue sky.

'Der Tag bricht!' exclaimed one of the men. I
looked towards the eastern heaven, but could discover no illumination which hinted at the approach
of day. At length the dawn really appeared,
brightening the blue of the eastern firmament;
at first it was a mere augmentation of cold
light, but by degrees it assumed a warmer tint.
The long uniform incline of the glacier being
passed, we reached the first eminences of snow
which heave like waves around the base of the
Jungfrau. This is the region of beauty in the
higher Alps—beauty pure and tender, out of which
emerges the savage scenery of the peaks. For the
healthy and the pure in heart these higher snow-
fields are consecrated ground.

The snow bosses were soon broken by chasms
deep and dark, which required tortuous winding on
our part to get round them. Having surmounted a
steep slope, we passed to some red and rotten rocks,
which required care on the part of those in front to
prevent the loose and slippery shingle from falling
upon those behind. We gained the ridge and
wound along it. High snow eminences now flanked
us to the left, and along the slope over which we
passed the *séracs* had shaken their frozen boulders. We tramped amid the knolls of the fallen
avalanches towards a white wall which, so far as we
could see, barred further progress. To our right

were noble chasms, blue and profound, torn into the
heart of the *névé* by the slow but resistless drag of
gravity on the descending snows. Meanwhile the
dawn had brightened into perfect day, and over
mountains and glaciers the gold and purple light
of the eastern heaven was liberally poured. We
had already caught sight of the peak of the
Jungfrau, rising behind an eminence and piercing
for fifty feet or so the rosy dawn. And many
another peak of stately altitude caught the blush,
while the shaded slopes were all of a beautiful azure,
being illuminated by the firmament alone. A large
segment of space enclosed between the Monk and
Trugberg was filled like a reservoir with purple
light. The world, in fact, seemed to worship, and
the flush of adoration was on every mountain-head.

Over the distant Italian Alps rose clouds of the
most fantastic forms, jutting forth into the heavens
like enormous trees, thrusting out umbrageous
branches which bloomed and glistened in the solar
rays. Along the whole southern heaven these fantastic masses were ranged close together, but still
perfectly isolated, until on reaching a certain altitude
they seemed to meet a region of wind which blew
their tops like streamers far away through the air.
Warmed and tinted by the morning sun, those unsub-
stantial masses rivalled in grandeur the mountains
themselves.

The final peak of the Jungfrau is now before us, and apparently *so* near ! But the mountaineer alone knows how delusive the impression of nearness often is in the Alps. To reach the slope which led up to the peak we must scale or round the barrier already spoken of. From the coping and the ledges of this beautiful wall hung long stalactites of ice, in some cases like inverted spears, with their sharp points free in air. In other cases, the icicles which descended from the overhanging top reached a projecting lower ledge, and stretched like a crystal railing from the one to the other. To the right of this barrier was a narrow gangway, from which the snow had not yet broken away so as to form a vertical or overhanging wall. It was one of those accidents which the mountains seldom fail to furnish, and on the existence of which the success of the climber entirely depends. Up this steep and narrow gangway we cut our steps, and a few minutes placed us safely at the bottom of the final pyramid of the Jungfrau.

From this point we could look down into the abyss of the Roththal, and certainly its wild environs seemed to justify the uses to which superstition has assigned the place. For here it is said the original demons of the mountains hold their orgies, and hither the spirits of the doubly-damned among men are sent to bear them company. The slope up which we had now to climb was turned towards the sun ; its aspect was a southern one, and its snows had been melted and recongealed to hard ice. The axe of Almer rung against the obdurate solid, and its fragments whirred past us with a weird-like sound to the abysses below. They suggested the fate which a false step might bring along with it. It is a practical tribute to the strength and skill of the Oberland guides that no disaster has hitherto occurred upon the peak of the Jungfrau.

The work upon this final ice-slope was long and heavy, and during this time the summit appeared to maintain its distance above us. We at length cleared the ice, and gained a stretch of snow which enabled us to treble our upward speed. Thence to some loose and shingly rocks, again to the snow, whence a sharp edge led directly up to the top. The exhilaration of success was here added to that derived from physical nature. On the top fluttered a little black flag, planted by our most recent predecessors. We reached it at 7.15 A.M., having accomplished the ascent from the Faulberg in six hours. The snow was flattened on either side of the apex so as to enable us all to stand upon it, and here we stood for some time, with all the magnificence of the Alps unrolled before us.

We may look upon those mountains again and again from a dozen different points of view, a

perennial glory surrounds them which associates with
every new prospect fresh impressions. I thought
I had scarcely ever seen the Alps to greater advantage. Hardly ever was their majesty more fully
revealed or more overpowering. The colouring of
the air contributed as much to the effect as the
grandeur of the masses on which that colouring fell.
A calm splendour overspread the mountains, softening the harshness of the outlines without detracting
from their strength. But half the interest of such
scenes is psychological; the soul takes the tint of surrounding nature, and in its turn becomes majestic.

And as I looked over this wondrous scene towards
Mont Blanc, the Grand Combin, the Dent Blanche,
the Weisshorn, the Dom, and the thousand lesser
peaks which seemed to join in celebration of the
risen day, I asked myself, as on previous occasions:
How was this colossal work performed? Who
chiselled these mighty and picturesque masses out of
a mere protuberance of the earth? And the answer
was at hand. Ever young, ever mighty—with the
vigour of a thousand worlds still within him—the
real sculptor was even then climbing up the eastern
sky. It was he who raised aloft the waters which
cut out these ravines; it was he who planted the
glaciers on the mountain-slopes, thus giving gravity
a plough to open out the valleys; and it is he who,
acting through the ages, will finally lay low these

mighty monuments, rolling them gradually sea-
ward—

Sowing the seeds of continents to be;

so that the people of an older earth may see mould
spread and corn wave over the hidden rocks which
at this moment bear the weight of the Jungfrau.[1]

[1] Eight years ago I was evidently a sun-worshipper; nor have I
yet lost the conviction of his ability to do all here ascribed to him.—
J. T., 1871

XVIII.

DEATH OF BENNEN ON THE HAUT DE CRY.

By PHILIP C. GOSSETT.

[On a March morning in 1864 I was returning to town from Chislehurst, when my attention was directed to an account of an Alpine disaster published in that day's ' Times.' No names were mentioned, and I commented rather severely on the rashness of trusting to mountain-snow so early in the year. On the following day I learned that my brave Bennen was one of the victims. Mr. P. C. Gossett wrote for the ' Alpine Journal ' a ' Narrative of the Accident,' which, through the obliging kindness of the author, I am enabled to publish here. Mr. Gossett was accompanied by his friend M. Boissonnet on the fatal day.]

On February 28, 1864, we left Sion with Bennen to mount the Haut de Cry. We started at 2.15 A.M. in a light carriage that brought us to the village of Ardon, distant six miles. We there met three men that were to accompany us as local guides and porters—Jean Joseph Nance, Frederic Rebot, who acted as my personal guide, and Auguste Bevard. We at once began to ascend on the right bank of the Lyzerne. The night was splendid, the sky cloudless, and the moon shining brightly. For about half an hour we went up through the vine-yards by a rather steep path, and then entered the valley of the Lyzerne, about 700 feet above the torrent. We here found a remarkably good path, gradually rising and leading towards the Col de Chéville. Having followed this path for about three hours, we struck off to the left, and began zigzagging up the mountain-side through a pine forest. We had passed what may be called the snow-line in winter a little above 2,000 feet. We had not ascended for more than a quarter of an hour in this pine forest before the snow got very deep and very soft. We had to change leader every five or six minutes, and even thus our progress was remarkably slow. We saw clearly that, should the snow be as soft above the fir region, we should have to give up the ascent. At 7 A.M. we reached a chalet, and stopped for about twenty minutes to rest and look at the sunrise on the Diablerets. On observing an aneroid, which we had brought with us, we found that we were at the height of about 7,000 feet: the temperature was −1° C.

The Haut de Cry has four *arêtes*, the first running towards the W., the second SE., the third E., and

the fourth NE. We were between the two last-
named *arêtes*. Our plan was to go up between them
to the foot of the peak, and mount it by the *arête*
running NE. As we had expected, the snow was in
much better state when once we were above the
woods. For some time we advanced pretty rapidly.
The peak was glistening before us, and the idea of
success put us in high spirits. Our good fortune did
not last long; we soon came to snow frozen on the
surface, and capable of bearing for a few steps and
then giving way. But this was nothing compared to
the trouble of pulling up through the pine wood, so
instead of making us grumble it only excited our
hilarity. Bennen was in a particularly good humour,
and laughed aloud at our combined efforts to get out
of the holes we every now and then made in the snow.
Judging from appearances, the snow-field over which
we were walking covered a gradually rising Alp.
We made a second observation with our aneroid, and
found, rather to our astonishment and dismay, that
we had only risen 1,000 feet in the last three hours.
It was 10 o'clock: we were at the height of about
8,000 feet; temperature $= -1\cdot 5$ C. During the
last half-hour we had found a little hard snow, so
we had all hope of success. Thinking we might
advance better on the *arête*, we took to it, and rose
along it for some time. It soon became cut up by
rocks, so we took to the snow again. It turned out

to be here hard frozen, so that we reached the
real foot of the peak without the slightest difficulty.
It was steeper than I had expected it would be,
judging from the valley of the Rhone. Bennen
looked at it with decided pleasure ; having completed
his survey, he proposed to take the eastern *arête*, as
in doing so we should gain at least two hours. Rebot
had been over this last-named *arête* in summer, and
was of Bennen's opinion. Two or three of the party
did not like the idea much, so there was a discussion
on the probable advantages and disadvantages of
the NE. and E. *arêtes*. We were losing time ; so
Bennen cut matters short by saying : 'Ich will der
Erste über die arête!' Thus saying, he made for
the E. *arête*; it looked very narrow, and, what was
worse, it was considerably cut up by high rocks, the
intervals between the teeth of the *arête* being filled
up with snow. To gain this *arête*, we had to go up
a steep snow-field, about 800 feet high, as well as I
remember. It was about 150 feet broad at the top,
and 400 or 500 at the bottom. It was a sort of
couloir on a large scale. During the ascent we sank
about one foot deep at every step. Bennen did not
seem to like the look of the snow very much. He
asked the local guides whether avalanches ever came
down this couloir, to which they answered that our
position was perfectly safe. We had mounted on the
northern side of the couloir, and having arrived at

150 feet from the top, we began crossing it on a horizontal curve, so as to gain the E. *arête*. The inflexion or dip of the couloir was slight, not above 25 feet, the inclination near 35°. We were walking in the following order: Bevard, Nance, Bennen, myself, Boissonnet, and Rebot. Having crossed over about three-quarters of the breadth of the couloir, the two leading men suddenly sank considerably above their waists. Bennen tightened the rope. The snow was too deep to think of getting out of the hole they had made, so they advanced one or two steps, dividing the snow with their bodies. Bennen turned round and told us he was afraid of starting an avalanche; we asked whether it would not be better to return and cross the couloir higher up. To this the three Ardon men opposed themselves; they mistook the proposed precaution for fear, and the two leading men continued their work. After three or four steps gained in the aforesaid manner, the snow became hard again. Bennen had not moved—he was evidently undecided what he should do; as soon, however, as he saw hard snow again, he advanced and crossed parallel to, but above, the furrow the Ardon men had made. Strange to say, the snow supported him. While he was passing I observed that the leader, Bevard, had about twenty feet of rope coiled round his shoulder. I of course at once told him to uncoil it and get on the *arête*, from which

he was not more than fifteen feet distant. Bennen
then told me to follow. I tried his steps, but sank
up to my waist in the very first. So I went through
the furrows, holding my elbows close to my body, so
as not to touch the sides. This furrow was about
twelve feet long, and, as the snow was good on the
other side, we had all come to the false conclusion
that the snow was accidentally softer there than elsewhere. Boissonet then advanced; he had made but
a few steps when we heard a deep, cutting sound.
The snow-field split in two about fourteen or fifteen
feet above us. The cleft was at first quite narrow,
not more than an inch broad. An awful silence
ensued; it lasted but a few seconds, and then it was
broken by Bennen's voice, 'Wir sind alle verloren.'
His words were slow and solemn, and those who
knew him felt what they really meant when spoken
by such a man as Bennen. They were his last
words. I drove my alpenstock into the snow, and
brought the weight of my body to bear on it; it
went in to within three inches of the top. I then
waited. It was an awful moment of suspense.
I turned my head towards Bennen to see whether
he had done the same thing. To my astonishment,
I saw him turn round, face the valley, and stretch
out both arms. The ground on which we stood
began to move slowly, and I felt the utter uselessness of any alpenstock. I soon sank up to my

shoulders and began descending backwards. From
this moment I saw nothing of what had happened
to the rest of the party. With a good deal of
trouble I succeeded in turning round. The speed
of the avalanche increased rapidly, and before long
I was covered up with snow and in utter darkness.
I was suffocating, when with a jerk I suddenly came
to the surface again. The rope had caught most
probably on a rock, and this was evidently the
moment when it broke. I was on a wave of the
avalanche, and saw it before me as I was carried
down. It was the most awful sight I ever witnessed. The head of the avalanche was already
at the spot where we had made our last halt. The
head alone was' preceded by a thick cloud of snow-
dust; the rest of the avalanche was clear. Around
me I heard the horrid hissing of the snow, and far
before me the thundering of the foremost part of
the avalanche. To prevent myself sinking again, I
made use of my arms much in the same way as
when swimming in a standing position. At last I
noticed that I was moving slower; then I saw the
pieces of snow in front of me stop at some yards'
distance; then the snow straight before me stopped,
and I heard on a large scale the same creaking
sound that is produced when a heavy cart passes
over hard-frozen snow in winter. I felt that I also
had stopped, and instantly threw up both arms to

protect my head in case I should again be covered
up. I had stopped, but the snow behind me was
still in motion; its pressure on my body was so
strong that I thought I should be crushed to death.
This tremendous pressure lasted but a short time,
and ceased as suddenly as it had begun. I was
then covered up by snow coming from behind me.
My first impulse was to try and uncover my head—
but this I could not do : the avalanche had frozen by
pressure the moment it stopped, and I was frozen
in. Whilst trying vainly to move my arms, I
suddenly became aware that the hands as far as the
wrist had the faculty of motion. The conclusion
was easy, they must be above the snow. I set to
work as well as I could; it was time, for I could
not have held out much longer. At last I saw a
faint glimmer of light. The crust above my head
was getting thinner, and it let a little air pass,
but I could not reach it any more with my
hands; the idea struck me that I might pierce
it with my breath. After several efforts I succeeded in doing so, and felt suddenly a rush of air
towards my mouth; I saw the sky again through
a little round hole. A dead silence reigned around
me ; I was so surprised to be still alive, and so
persuaded at the first moment that none of my
fellow-sufferers had survived, that I did not even
think of shouting for them. I then made vain

efforts to extricate my arms, but found it impossible; the most I could do was to join the ends
of my fingers, but they could not reach the snow
any longer. After a few minutes I heard a man
shouting: what a relief it was to know that I was
not the sole survivor! to know that perhaps he was
not frozen in and could come to my assistance! I
answered; the voice approached, but seemed uncertain where to go, and yet it was now quite near.
A sudden exclamation of surprise! Rebot had seen
my hands. He cleared my head in an instant, and
was about to try and cut me out completely, when
I saw a foot above the snow, and so near to me that
I could touch it with my arms, although they were
not quite free yet. I at once tried to move the
foot; it was my poor friend's. A pang of agony
shot through me as I saw that the foot did not
move. Poor Boissonnet had lost sensation, and was
perhaps already dead. Rebot did his best: after
some time he wished me to help him, so he freed
my arms a little more, so that I could make use of
them. I could do but little, for Rebot had torn the
axe from my shoulder as soon as he had cleared my
head (I generally carry an axe separate from my
alpenstock—the blade tied to the belt, and the
handle attached to the left shoulder). Before coming to me Rebot had helped Nance out of the snow;
he was lying nearly horizontally, and was not much

JOHANN JOSEPH BENNEN.

covered over. Nance found Bevard, who was upright in the snow, but covered up to the head.
After about twenty minutes the two last-named
guides came up. I was at length taken out; the
snow had to be ·cut with the axe down to my feet
before I could be pulled out. A few minutes after
1 o'clock P.M. we came to my poor friend's face. . . .
I wished the body to be taken out completely, but
nothing could induce the three guides to work any
longer, from the moment they saw that it was too
late to save him. I acknowledge that they were
nearly as incapable of doing anything as I was.
When I was taken out of the snow the cord had to
be cut. We tried the end going towards Bennen,
but could not move it; it went nearly straight
down and showed us that there was the grave of the
bravest guide the Valais ever had, and ever will
have. The cold had done its work on us ; we could
stand it no longer, and began the descent. We
followed the frozen avalanche for about twenty-five
minutes, that being the easiest way of progressing,
and then took the track we had made in the
morning ; in five hours we reached Ardon.

I have purposely put apart the details I have been
asked to give on certain points.

1. The avalanche consisted only of snow; the
upper stratum was eleven days old. At the moment,
the avalanche started it was about twelve o'clock,

probably a few minutes before. The temperature was then above freezing point, and we were within 300 or 350 feet from the summit. The snow was thawing, and the whole snow-field in a state of uncertain equilibrium. By cutting through the snow at the top of the couloir we cut one of the main points by which the snow of the two different layers held together; what led us into the error was, as I have before said, the fact that the snow was quite hard in some places, and quite soft in others. The avalanche may have taken a minute to descend; I can give no correct estimation on this point. We fell between 1,900 and 1,960 feet, the head of the avalanche going 800 feet lower.

2. The rope was in my opinion the cause of my poor friend's as well as of Bennen's death. The following facts may prove it: At the moment the avalanche started the first and last guides merely held the rope; Bennen had not seen the use of a rope at all, so we had been less strict than we should otherwise have been in its use. During the descent the rope caught, probably on a rock below the surface. This happened between Bennen and Nance, that is to say between the second and third man in the marching line. Nance told me afterwards that this was the worst part of the descent; he had the pressure of the snow on his body, whilst the rope nearly cut him in two. I believe that it was at this

moment that Bennen and Boissonnet lost their
upright position, owing to the pressure of snow on
their backs. Nance also lost his position, but was
fortunate in being thrown out horizontally, and
that almost on the surface of the avalanche. I was
between Bennen and Boissonnet, but not tied to
the rope, as I had iron rings to my belt through
which the cord ran. Rebot, who was last in the
line, was thrown clean out of the avalanche; he
was carried during the descent towards one of the
sides of the stream. He was the only one of us who
escaped unhurt. Thus, when we stopped in our
descent, two only were *tied* to the rope—Boissonnet
and Bennen—the very two who perished.

3. The congealing of the snow happened by pressure. The fore part of the avalanche stopped first,
and the rest was forced against it. The circumstance
I can least understand is the sudden fall in the temperature of the air after the accident. I can give no
estimate of it, but it was intense.

4. The bruises Bevard, Nance, and I sustained
were slight, but our feet were severely frost-bitten.
Bennen has been accused of rashness in this unfortunate accident. It is not the case. He was misled
by the total difference of the state of snow in a winter
ascent from what is to be met with in summer.

I have been recently favoured with a letter from Mr. Gossett, from which the following is an extract :

Berne: March 17, 1871.

' Bennen's body was found with great difficulty the third day after Boissonnet was found. The cord-end had been covered up with snow. The curé d'Ardon informed me that poor Bennen was found eight feet under the snow, in a horizontal position, the head facing the valley of the Lyzerne. His watch had been wrenched from the chain, probably when the cord broke ; the chain, however, remained attached to his waistcoat. Three years ago I met one of my Ardon guides ; he told me that Bennen's watch had been found by a shepherd seven months after the accident. This shepherd had been one of the party who went up to look for Bennen ; during the following summer he had watched the melting of the avalanche. When mounted, the watch obeyed. This reminds me of your fall on the Morteratsch glacier.[1]

' I know you were very much attached to Bennen ; the same was the case with him in regard to you. An hour before his death the Matterhorn showed its black head over one of the *arêtes* of the Haut de Cry. I asked Bennen whether he thought it would ever be ascended. His answer was a decided " Yes " ;

[1] See Chapter XIX.

but he added, alluding to your last attack on the mountain, "Wir waren fünf; der Professor und ich stimmten für Vorwärts; die drei andern stimmten dagegen."

'There is one circumstance in reference to my fall with the avalanche of the Haut de Cry that I am utterly unable to understand : I mean what physical phenomena took place when the avalanche stopped and froze. It stopped because in its progress downwards the broad couloir down which it was going got narrower, and the mass of snow could not pass. It froze because the successive portions of the body of the avalanche became compressed against the head, which latter had come to a stop. When the layer in which I was stopped, the pressure on my body was enormous—so great, in fact, that I expected I should be crushed flat. This pressure ceased *suddenly* : I know it, for the atrocious pains it was causing ceased *suddenly* too. What happened during that interval ? '

[Bennen was well acquainted with winter snow ; but no man of his temper, and in his position, would place himself in direct opposition to local guides, whose knowledge of the mountain must have been superior to his own.]

XIX.

ACCIDENT ON THE PIZ MORTERATSCH.

WHILE staying at Pontresina in 1864 I joined Mr. Hutchinson, and Mr. Lee-Warner, of Rugby, in a memorable expedition up the Piz Morteratsch. This is a very noble mountain, and, as we thought, safe and easy to ascend. The resolute Jenni, by far the boldest man in Pontresina, was my guide; while Walter, the official *guide chef*, was taken by my companions. With a dubious sky overhead, we started on the morning of July 30, a little after four A.M. There is rarely much talk at the beginning of a mountain excursion: you are either sleepy or solemn so early in the day. Silently we passed through the pine woods of the beautiful Rosegg valley, watching anxiously at intervals the play of the clouds around the adjacent heights. At one place a spring gushed from the valley-bottom, as clear and almost as copious as that which pours out the full-formed river Albula. The traces of ancient glaciers were present everywhere, the valley being thickly covered with the rubbish which the ice had left behind. An ancient moraine, so large that in England it might take rank as a mountain, forms a barrier across the upper valley. Once probably it was the dam of a lake, but it is now cut through by the river which rushes from the Rosegg glacier. These works of the ancient ice are to the mind what a distant horizon is to the eye. They give to the imagination both pleasure and repose.

The morning, as I have said, looked threatening, but the wind was good; by degrees the cloud-scowl relaxed, and broader patches of blue became visible above us. We called at the Rosegg chalets, and had some milk. We afterwards wound round a shoulder of the hill, at times upon the moraine of the glacier, and at times upon the adjacent grass slope; then over shingly inclines, covered with the shot rubbish of the heights. Two ways were now open to us, the one easy but circuitous, the other stiff but short. Walter was for the former, and Jenni for the latter, their respective choices being characteristic of the two men. To my satisfaction Jenni prevailed, and we scaled the steep and slippery rocks. At the top of them we found ourselves upon the rim of an extended snow-field. Our rope was here exhibited, and we were bound by it to a common destiny. In those higher regions the snow-fields

P

show a beauty and a purity of which persons who linger low down have no notion. We crossed crevasses and bergschrunds, mounted vast snow-basses, and doubled round walls of ice with long stalactites pendent from their eaves. One by one the eminences were surmounted. The crowning rock was attained at half-past twelve. On it we uncorked a bottle of champagne ; mixed with the pure snow of the mountain, it formed a beverage, and was enjoyed with a gusto, which the sybarite of the city could neither imitate nor share.

We spent about an hour upon the warm gneiss-blocks on the top. Veils of cloud screened us at intervals from the sun, and then we felt the keenness of the air ; but in general we were cheered and comforted by the solar light and warmth. The shiftings of the atmosphere were wonderful. The white peaks were draped with opalescent clouds which never lingered for two consecutive minutes in the same position. Clouds differ widely from each other in point of beauty, but I had hardly seen them more beautiful than they appeared to-day, while the succession of surprises experienced through their changes were such as rarely fall to the lot even of an experienced mountaineer.

These clouds are for the most part produced by the chilling of the air through its own expansion. When thus chilled, the aqueous vapour diffused

through it, which is previously unseen, is precipitated in visible particles. Every particle of the
cloud has consumed in its formation a little polyhedron of vapour, and a moment's reflection will
make it clear that the size of the cloud-particles
must depend, not only on the size of the vapour
polyhedron, but on the relation of the density of the
vapour to that of its liquid. If the vapour were
light and the liquid heavy, other things being
equal, the cloud-particle would be *smaller* than if
the vapour were heavy and the liquid light. There
would evidently be more *shrinkage* in the one case
than in the other. Now there are various liquids
whose weight is not greater than that of water,
while the weight of their vapours, bulk for bulk, is
five or six times that of aqueous vapour. When
those heavy vapours are precipitated as clouds,
which is easily done artificially, their particles are
found to be far coarser than those of an aqueous
cloud. Indeed water is without a parallel in this
particular. Its vapour is the lightest of all vapours,
and to this fact the soft and tender beauty of the
clouds of our atmosphere is mainly due.[1]

After an hour's halt upon the summit the descent
began. Jenni is the most daring man and powerful character among the guides of Pontresina.
The manner in which he bears down all the others

[1] Chapter V, p. 405, is devoted to ' Clouds.' See also note, p. 82.

in conversation, and imposes his own will upon
them, shows that he is the dictator of the place.
He is a large and rather an ugly man, and his
progress up hill, though resistless, is slow. He had
repeatedly expressed a wish to make an excursion
with me, and on this occasion he may have desired
to show us what he could do upon the mountains.
He accomplished two daring things—the one success-
fully, while the other was within a hair's-breadth
of a very shocking issue.

In descending we went straight down upon a
bergschrund, which had compelled us to make a
circuit in coming up. This particular kind of
fissure is formed by the lower portion of a snow-
slope falling away from the upper, a crevasse being
thus formed between both, which often surrounds
the mountain as a fosse of terrible depth. Walter
was the first of our party, and Jenni was the last.
It was quite evident that the leader hesitated to
cross the chasm ; but Jenni came forward, and half
by expostulation, half by command, caused him to
sit down on the snow at some height above the
fissure. I think, moreover, he helped him with a
shove. At all events, the slope was so steep that
the guide shot down it with an impetus sufficient to
carry him clear over the schrund. We all afterwards shot the chasm in this pleasant way. Jenni
was behind. Deviating from our track, he

deliberately chose the widest part of the chasm, and shot over it, lumbering like behemoth down the snow-slope at the other side. It was an illustration of that practical knowledge which long residence among the mountains can alone impart, and in the possession of which our best English climbers fall far behind their guides.

The remaining steep slopes were also descended by glissade, and we afterwards marched cheerily over the gentler inclines. We had ascended by the Rosegg glacier, and now we wished to descend upon the Morteratsch glacier and make it our highway home.

We reached the point at which it was necessary to quit our morning's track, and immediately afterwards got upon some steep rocks, rendered slippery here and there by the water which trickled over them. To our right was a broad couloir, filled with snow, which had been melted and refrozen, so as to expose a steeply sloping wall of ice. We were tied together in the following order: Jenni led, I came next, then Mr. Hutchinson, a practised mountaineer, then Mr. Lee-Warner, and last of all the guide Walter. Lee-Warner had had but little experience of the higher Alps, and he was placed in front of Walter, so that any false step on his part might be instantly checked.

After descending the rocks for a time Jenni turned

and asked me whether I thought them or the ice-slope
the better track. I pronounced without hesitation·in
favour of the rocks, but he seemed to misunderstand
me, and turned towards the couloir.   I stopped
him at the edge of it, and said, 'Jenni, you know
where you are going; the slope is pure ice.' He
replied, 'I know it; but the ice is quite bare for a
few yards only. Across this exposed portion I will
cut steps, and then the snow which covers the ice
will give us a footing.' He cut the steps, reached
the snow, and descended carefully along it, all following him, apparently in good order. After some
time he stopped, turned, and looked upwards at the
last three men. 'Keep carefully in the steps, gentlemen,' he said; 'a false step here might detach an
avalanche.' The word was scarcely uttered when I
heard the sound of a fall behind me, then a rush,
and in a moment my two friends and their guide, all
apparently entangled together, whirred past me. I
suddenly planted myself to resist their shock, but in
an instant I was in their wake, for their impetus was
irresistible. A moment afterwards Jenni was whirled
away, and thus, in the twinkling of an eye, all five
of us found ourselves riding downwards with un-
controllable speed on the back of an avalanche
which a single slip had originated.

Previous to stepping on the slope, I had, according to habit, made clear to my mind what was to be

done in case of mishap; and accordingly, when over-thrown, I turned promptly on my face and drove my bâton through the moving snow, and into the ice underneath. No time, however, was allowed for the break's action; for I had held it firmly thus for a few seconds only, when I came into collision with some obstacle and was rudely tossed through the air, Jenni at the same time being shot down upon me. Both of us here lost our bâtons. We had been carried over a crevasse, had hit its lower edge, and, instead of dropping into it, were pitched by our great velo-city far beyond it. I was quite bewildered for a moment, but immediately righted myself, and could see the men in front of me half buried in the snow, and jolted from side to side by the ruts among which we were passing. Suddenly I saw them tumbled over by a lurch of the avalanche, and immediately afterwards found myself imitating their motion. This was caused by a second crevasse. Jenni knew of its existence and plunged, he told me, right into it—a brave act, but for the time unavailing. By jumping into the chasm he thought a strain might be put upon the rope sufficient to check the motion. But, though over thirteen stone in weight, he was violently jerked out of the fissure and almost squeezed to death by the pressure of the rope.

A long slope was below us, which led directly down-wards to a brow where the glacier fell precipitously.

At the base of the declivity the ice was cut by a series of profound chasms, towards which we were rapidly borne. The three foremost men rode upon the forehead of the avalanche, and were at times almost wholly immersed in the snow; but the moving layer was thinner behind, and Jenni rose incessantly and with desperate energy drove his feet into the firmer substance underneath. His voice, shouting 'Halt! Herr Jesus, halt!' was the only one heard during the descent. A kind of condensed memory, such as that described by people who have narrowly escaped drowning, took possession of me, and my power of reasoning remained intact. I thought of Bennen on the Haut de Cry, and muttered, 'It is now my turn.' Then I coolly scanned the men in front of me, and reflected that, if their *vis viva* was the only thing to be neutralised, Jenni and myself could stop them; but to arrest both them and the mass of snow in which they were caught was hopeless. I experienced no intolerable dread. In fact, the start was too sudden and the excitement of the rush too great to permit of the development of terror.

Looking in advance, I noticed that the slope, for a short distance, became less steep, and then fell as before. 'Now or never we must be brought to rest.' The speed visibly slackened, and I thought we were saved. But the momentum had been too great: the avalanche crossed the brow and in part regained its motion. Here Hutchinson threw his arm round his friend, all hope being extinguished, while I grasped my belt and struggled to free myself. Finding this difficult, from the tossing, I sullenly resumed the strain upon the rope. Destiny had so related the downward impetus to Jenni's pull as to give the latter a slight advantage, and the whole question was whether the opposing force would have sufficient time to act. This was also arranged in our favour, for we came to rest so near the brow that two or three seconds of our average motion of descent must have carried us over. Had this occurred, we should have fallen into the chasms, and been covered up by the tail of the avalanche. Hutchinson emerged from the snow with his fore-head bleeding, but the wound was superficial; Jenni had a bit of flesh removed from his hand by collision against a stone; the pressure of the rope had left black welts on my arms; and we all experienced a tingling sensation over the hands, like that pro-duced by incipient frostbite, which continued for several days. This was all. I found a portion of my watch-chain hanging round my neck, another portion in my pocket; the watch was gone.

This happened on the 30th of July. Two days afterwards I went to Italy, and remained there for ten or twelve days. On the 16th of August, being again at Pontresina, I made on that day an

expedition in search of the lost watch. Both the
guides and myself thought the sun's heat might
melt the snow above it, and I inferred that if its
back should happen to be uppermost the slight
absorbent power of gold for the solar rays would
prevent the watch from sinking as a stone sinks
under like circumstances. The watch would thus
be brought quite to the surface; and, although a
small object, it might possibly be seen from some
distance.   Five friends accompanied me up the
Morteratsch glacier.   One of them was the late
Mr. North, member for Hastings, a most lovable
man. He was then sixty-four years of age, but he
exhibited a courage and collectedness, and indeed a
delight, in the wild savagery of the crevasses which
were perfectly admirable.

Two only of the party, both competent mountaineers, accompanied me to the track of our glissade, but none of us ventured on the ice where it
originated. Just before stepping upon the snow,
a stone some tons in weight, detached by the sun
from the heights above us, came rushing down the
line of our descent. Its leaps became more and more
impetuous, and on reaching the brow near which we
had been brought to rest it bounded through the
air, and with a single spring reached the lower
glacier, raising a cloud of ice-dust. Some fragments of rope found upon the snow assured us that

we were upon the exact track of the avalanche, and then the search commenced. It had not continued twenty minutes when a cheer from one of the guides —Christian Michel of Grindelwald—announced the discovery of the watch. It had been brought to the surface in the manner surmised, and on examination seemed to be dry and uninjured. I noticed, moreover, that the position of the hands indicated that it had only run down beneath the snow. I wound it up, hardly hoping, however, to find it capable of responding. But it showed instant signs of animation. It had remained eighteen days in the avalanche, but the application of its key at once restored it to action, and it has gone with unvarying regularity ever since.

Mr. Hutchinson has published the following note of the accident in the 'Alpine Journal':

' As one of the party concerned in the accident on the Piz Morteratsch last July, I trust I shall not be thought presumptuous in bearing my testimony to the entire accuracy of Professor Tyndall's account. I can add no facts of any importance to those there mentioned, unless it be that we estimated the distance down which we were carried at fully 1,000 feet—a conclusion which, Mr. Tyndall tells me, was confirmed by his subsequent visit to the spot. The angle of the slope we did not measure, nor can I

give the time of our descent with any accuracy; it
seemed to me a lifetime. From the moment that
the snow cracked, Jenni behaved with the greatest
coolness and courage. But he ought not to have
taken us down the ice-slope so late in the day—it
was then nearly half-past two o'clock—and that
after a warning word from Professor Tyndall and
myself. Of Walter's conduct the less said the
better; our opinion of his courage was not raised
by this trial of it.'

[Until Mr. Gossett's letter reached me a few days
ago I was not aware of the singular likeness between
the loss of Bennen's watch and of my own.—April
1871.]

THE GORGE OF PFEFFERS (SHOWING EROSIVE ACTION).

XX.

ALPINE SCULPTURE.

To the physical geologist the conformation of the Alps, and of mountain-regions generally, constitutes one of the most interesting problems of the present day. To account for this conformation, two hypotheses have been advanced, which may be respectively named the hypothesis of *fracture* and the hypothesis of *erosion*. Those who adopt the former maintain that the forces by which the Alps were elevated produced fissures in the earth's crust, and that the valleys of the Alps are the tracks of these fissures. Those who hold the latter hypothesis maintain that the valleys have been cut out by the action of ice and water, the mountains themselves being the residual forms of this grand sculpture. To the erosive action here indicated must be added that due to the atmosphere (the severance and detachment of rocks by rain and frost), as affecting the forms of the more exposed and elevated peaks.

I had heard it stated that the Via Mala was a striking illustration of the fissure theory—that the profound chasm thus named, and through which the

Hinter-Rhein now flows, could be nothing else than a crack in the earth's crust.   To the Via Mala I therefore went in 1864 to instruct myself by actual observation upon the point in question.

The gorge commences about a quarter of an hour above Tusis; and, on entering it, the first conclusion is that it must be a fissure.   This conclusion in my case was modified as I advanced. Some distance up the gorge I found upon the slopes to my right quantities of rolled stones, evidently rounded by water-action.   Still further up, and just before reaching the first bridge which spans the chasm, I found more rolled stones, associated with sand and gravel.   Through this mass of detritus, fortunately, a vertical cutting had been made, which exhibited a section showing perfect stratification. There was no agency in the place to roll these stones, and to deposit these alternating layers of sand and pebbles, but the river which now rushes some hundreds of feet below them.   At one period of the Via Mala's history the river must have run at this high level.   Other evidences of water-action soon revealed themselves.   From the parapet of the first bridge I could see the solid rock 200 feet above the bed of the river scooped and eroded.

It is stated in the guide-books that the river, which usually runs along the bottom of the gorge, has been known almost to fill it during violent thunderstorms; and it may be urged that the marks of erosion which the sides of the chasm exhibit are due to those occasional floods. In reply to this, it may be stated that even the existence of such floods is not well authenticated, and that if the supposition were true, it would be an additional argument in favour of the cutting power of the river. For if floods operating at rare intervals could thus erode the rock, the same agency, acting without ceasing upon the river's bed, must certainly be competent to excavate it.

I proceeded upwards, and from a point near another bridge (which of them I did not note) had a fine view of a portion of the gorge. The river here runs at the bottom of a cleft of profound depth, but so narrow that it might be leaped across. That this cleft must be a crack is the impression first produced; but a brief inspection suffices to prove that it has been cut by the river. From top to bottom we have the unmistakable marks of erosion. This cleft was best seen by looking downwards from a point near the bridge; but looking upwards from the bridge itself, the evidence of aqueous erosion was equally convincing.

The character of the erosion depends upon the rock as well as upon the river. The action of water upon some rocks is almost purely mechanical; they are simply ground away or detached in sensible

masses. In other cases the action is chemical as well
as mechanical. Water, in passing over limestone,
charges itself with carbonate of lime without damage to its transparency; the rock is *dissolved* in the
water; and the gorges cut by water in such rocks
often resemble those cut in the ice of glaciers by
glacier streams. To the solubility of limestone is
probably to be ascribed the fantastic forms which
peaks of this rock usually assume, and also the
grottos and caverns which interpenetrate limestone
formations. A rock capable of being thus dissolved
will expose a smooth surface after the water has
quitted it; and in the case of the Via Mala it is the
polish of the surfaces, and also the curved hollows
scooped in the sides of the gorge, which assure us
that the chasm has been the work of the river.

About four miles from Tusis, and not far from
the little village of Zillis, the Via Mala opens into a
plain bounded by high terraces, evidently cut by
water. It occurred to me the moment I saw it that
the plain had been the bed of an ancient lake; and
a farmer, who was my temporary companion, immediately informed me that such was the tradition of
the neighbourhood. This man conversed with intelligence, and as I drew his attention to the rolled
stones, which rest not only above the river, but above
the road, and inferred that the river must have been
there to have rolled those stones, he saw the force of

the evidence perfectly. In fact, in former times, and subsequent to the retreat of the great glaciers, a rocky barrier crossed the valley at this place, damming the river which came from the glaciers higher up. A lake was thus formed which poured its waters over the barrier. Two actions were here at work, both tending to obliterate the lake—the raising of its bed by the deposition of detritus, and the cutting of its dam by the river. In process of time the cut deepened into the Via Mala; the lake was drained, and the river now flows in a definite channel through the plain which its waters once totally covered.

From Tusis I crossed to Tiefenkasten by the Schien Pass, and thence over the Julier Pass to Pontresina. There are three or four ancient lake-beds between Tiefenkasten and the summit of the Julier. They are all of the same type—a more or less broad and level valley-bottom, with a barrier in front through which the river has cut a passage, the drainage of the lake being the consequence. These lakes are sometimes dammed by barriers of rock, sometimes by the moraines of ancient glaciers.

An example of this latter kind occurs in the Rosegg valley, about twenty minutes below the end of the Rosegg glacier, and about an hour from Pontresina. The valley here is crossed by a pine-covered moraine of the noblest dimensions: in the neighbourhood of

Q

London it might be called a mountain. That it is
a moraine, the inspection of it from a point on the
Surlei slopes above it will convince any person possessing an educated eye. Where, moreover, the interior of the mound is exposed, it exhibits moraine-
matter—detritus pulverised by the ice, with boulders
entangled in it. It stretched quite across the valley,
and at one time dammed the river up. But now the
barrier is cut through, the stream having about one-
fourth of the moraine to its right, and the remaining
three-fourths to its left. Other moraines of a more
resisting character hold their ground as barriers to
the present day. In the Val di Campo, for example,
about three-quarters of an hour from Pisciadello,
there is a moraine composed of large boulders, which
interrupt the course of a river and compel the water
to fall over them in cascades. They have in great
part resisted its action since the retreat of the
ancient glacier which formed the moraine. Behind
the moraine is a lake-bed, now converted into a
meadow, which is quite level, and rests on a deep
layer of mould.

At Pontresina a very fine and instructive gorge is
to be seen. The river from the Morteratsch glacier
rushes through a deep and narrow chasm which is
spanned at one place by a stone bridge. The rock
is not of a character to preserve smooth polishing;
but the larger features of water-action are perfectly

evident from top to bottom. Those features are in
part visible from the bridge, but still better from
a point a little distance from the bridge in the
direction of the upper village of Pontresina. The
hollowing out of the rock by the eddies of the water
is here quite manifest. A few minutes' walk upwards brings us to the end of the gorge; and behind
it we have the usual indications of an ancient lake,
and terraces of distinct water origin.

From this position the genesis of the gorge is
clearly revealed. After the retreat of the ancient
glacier, a transverse ridge of comparatively resisting
material crossed the valley at this place. Over the
lowest part of this ridge the river flowed, rushing
steeply down to join at the bottom of the slope the
stream which issued from the Rosegg glacier. On
this incline the water became a powerful eroding
agent, and finally cut its channel to its present depth.

Geological writers of reputation assume at this
place the existence of a fissure, the ' washing out '
of which resulted in the formation of the gorge.
Now no examination of the bed of the river ever
proved the existence of this fissure; and it is certain
that water can cut a channel through unfissured rock
—that cases of deep cutting can be pointed out
where the clean bed of the stream is exposed, the rock
which forms the floor of the river not exhibiting a
trace of fissure. An example of this kind occurs near

the Bernina Gasthaus, about two hours from Pontresina. A little way below the junction of the two
streams from the Bernina Pass and the Heuthal the
river flows through a channel cut by itself, and 20
or 30 feet in depth. At some places the river-bed
is covered with rolled stones; at other places it is
bare, but shows no trace of fissure. The abstract
power of water (if I may use the term) to cut
through rock is demonstrated by such instances.
But if water be competent to form a gorge without
the aid of a fissure, why assume the existence of such
in cases like that at Pontresina? It seems far more
philosophical to accept the simple and impressive
history written on the walls of those gorges by the
agent which produced them.

Numerous cases might be pointed out, varying
in magnitude, but all identical in kind, of barriers
which crossed valleys and formed lakes having been
cut through by rivers, narrow gorges being the consequence. One of the most famous examples of
this kind is the Finsteraarschlucht in the valley of
Hasli. Here the ridge called the Kirchet seems
split across, and the river Aar rushes through the
fissure. Behind the barrier we have the meadows
and pastures of Imhof resting on the sediment of
an ancient lake. Were this an isolated case, one
might reasonably conclude that the Finisteraar-
schlucht was produced by an earthquake, as some

suppose it to have been ; but when we find it to be
a single sample of actions which are frequent in
the Alps—when probably a hundred cases of the
same kind, though different in magnitude, can be
pointed out—it seems quite unphilosophical to
assume that in each particular case an earthquake
was at hand to form a channel for the river. As in
the case of the barrier at Pontresina, the Kirchet,
after the retreat of the Aar glacier, dammed the
waters flowing from it, thus forming a lake, on the
bed of which now stands the village of Imhof.
Over this barrier the Aar tumbled towards Meyringen, cutting, as the centuries passed, its bed ever
deeper, until finally it became deep enough to drain
the lake, leaving in its place the alluvial plain,
through which the river now flows in a definite
channel.[1]

But the broad view taken by the advocates of the
fracture theory is, that the valleys are the tracks of
primeval fissures produced by the upheaval of the
land, and the cracks across the barriers to which
I have referred are in reality portions of the
great cracks which formed the valleys. Such an
argument, however, would virtually concede the
theory of erosion as applied to the *valleys* of the
Alps. The narrow gorges, often not more than
twenty or thirty feet across, sometimes even

[1] For further observations see p. 258.

narrower, frequently occur at the bottom of broad
valleys.   Such fissures might enter into the list of
accidents which gave direction to the real erosive
agents which scooped the valley out ; but the formation of the valley, as it now exists, could no
more be ascribed to it than the motion of a railway
train could be ascribed to the finger of the engineer
which turns on the steam.

These deep gorges occur, I believe, for the most
part in limestone strata ; and the effects which the
merest driblet of water can produce on such rocks
are quite astonishing.   It is not uncommon to meet
chasms of considerable depth produced by small
streams the beds of which are dry for a large portion
of the year.   Right and left of the larger gorges
such secondary chasms are usually to be found.
The idea of *time* must, I think, be more and more
included in our reasonings on these phenomena.
Happily, the marks which the rivers have, in most
cases, left behind them, and which refer, geologi-
cally considered, to actions of yesterday, give us
ground and courage to conceive what may be effected in geologic periods.   Thus the modern portion of the Via Mala throws light upon the whole.
Near Bergün, in the valley of the Albula, there is
also a little Via Mala, which is not less significant
than the great one.   The river flows here through
a profound limestone gorge ; but to the very edges

of the gorge we have the evidences of erosion.   The
most striking illustration of water-action upon
limestone rock which I have ever witnessed is, I
think, furnished by the gorge at Pfäffers.   Here the
traveller passes along the side of the chasm midway
between top and bottom.   Whichever way he looks,
backwards or forwards, upwards or downwards, towards the sky or towards the river, he meets everywhere the irresistible and impressive evidence that
this wonderful fissure has been sawn through the
mountain by the waters of the Tamina.

I have thus far confined myself to the consideration of the gorges formed by the cutting through
of the rock-barriers which frequently cross the
valleys of the Alps; as far as I have examined them
they are the work of erosion.   But the larger
question still remains, To what action are we to
ascribe the formation of the valleys themselves?
This question includes that of the formation of the
mountain-ridges, for were the valleys wholly filled,
the ridges would disappear.   Possibly no answer
can be given to this question which is not beset
with more or less of difficulty.   Special localities
might be found which would seem to contradict
every solution which refers the conformation of the
Alps to the operation of a single cause.

Still the Alps present features of a character sufficiently definite to bring the question of their origin

within the sphere of close reasoning. That they were in whole or in part once beneath the sea will not be disputed; for they are in great part composed of sedimentary rocks which required a sea to form them. Their present elevation above the sea is due to one of those local changes in the shape of the earth which have been of frequent occurrence throughout geologic time, and which in some cases have depressed the land, and in others caused the sea-bottom to protrude beyond its surface. Considering the inelastic character of its materials, the protuberance of the Alps could hardly have been pushed out without dislocation and fracture; and this conclusion gains in probability when we consider the foldings, contortions, and even reversals in position of the strata in many parts of the Alps. Such changes in the position of beds which were once horizontal could not have been effected without dislocation. Fissures would be produced by these changes; and such fissures, the advocates of the fracture theory contend, mark the positions of the valleys of the Alps.

Imagination is necessary to the man of science, and we could not reason on our present subject without the power of presenting mentally a picture of the earth's crust cracked and fissured by the forces which produced its upheaval. Imagination, however, must be strictly checked by reason and by observation. That fractures occurred cannot, I think, be doubted, but that the valleys of the Alps are thus formed is a conclusion not at all involved in the admission of dislocations. I never met with a precise statement of the manner in which the advocates of the fissure theory suppose the forces to have acted—whether they assume a general elevation of the region, or a local elevation of distinct ridges ; or whether they assume local subsidences after a general elevation, or whether they would superpose upon the general upheaval minor and local upheavals.

In the absence of any distinct statement, I will assume the elevation to be general—that a swelling out of the earth's crust occurred here, sufficient to place the most prominent portions of the protuberance three miles above the sea-level. To fix the ideas, let us consider a circular portion of the crust, say one hundred miles in diameter, and let us suppose, in the first instance, the circumference of this circle to remain fixed, and that the elevation was confined to the space within it. The upheaval would throw the crust into a state of strain ; and, if it were inflexible, the strain must be relieved by fracture. Crevasses would thus intersect the crust. Let us now enquire what proportion the area of these open fissures is likely to bear to the area of the unfissured crust. An approximate

answer is all that is here required; for the problem
is of such a character as to render minute precision
unnecessary.

No one, I think, would affirm that the area of the
fissures would be one-hundredth the area of the land.
For let us consider the strain upon a single line
drawn over the summit of the protuberance from
a point on its rim to a point opposite. Regarding
the protuberance as a spherical swelling, the length
of the arc corresponding to a chord of 100 miles
and a versed sine of 3 miles is 100·24 miles; consequently the surface to reach its new position must
stretch 0·24 of a mile, or be broken. A fissure or a
number of cracks with this total width would relieve
the strain; that is to say, the sum of the widths of
all the cracks over the length of 100 miles would be
420 yards. If, instead of comparing the width of the
fissures with the length of the lines of tension, we
compared their areas with the area of the unfissured
land, we should of course find the proportion much
less. These considerations will help the imagination to realise what a small ratio the area of the
open fissures must bear to the unfissured crust.
They enable us to say, for example, that to assume
the area of the fissures to be one-tenth of the area
of the land would be quite absurd, while that the
area of the fissures could be one-half or more than
one-half that of the land would be in a proportionate

degree unthinkable. If we suppose the elevation
to be due to the shrinking or subsidence of the
land all round our assumed circle, we arrive equally
at the conclusion that the area of the open fissures
would be altogether insignificant as compared with
that of the unfissured crust.

To those who have seen them from a commanding
elevation, it is needless to say that the Alps themselves bear no sort of resemblance to the picture
which this theory presents to us. Instead of deep
cracks with approximately vertical walls, we have
ridges before us running into peaks, and gradually
sloping to form valleys. Instead of a fissured crust,
we have a state of things closely resembling the
surface of the ocean when agitated by a storm.
The valleys, instead of being much narrower than
the ridges, occupy the greater space. A plaster
cast of the Alps turned upside down, so as to invert
the elevations and depressions, would exhibit blunter
and broader mountains, with narrower valleys between them, than the present ones. The valleys
that exist cannot, I think, with any correctness of
language be called fissures. It may be urged that
they originated in fissures: but even this is un-
proved, and, were it proved, would still make the
fissures play the subordinate part of giving direction
to the agents which are to be regarded as the real
sculptors of the Alps.

The fracture theory, then, if it regards the elevation of the Alps as due to the operation of a force
acting throughout the entire region, is, in my
opinion, utterly incompetent to account for the
conformation of the country. If, on the other hand,
we are compelled to resort to local disturbances,
the manipulation of the earth's crust necessary to
obtain the valleys and the mountains will, I imagine, bring the difficulties of the theory into very
strong relief. Indeed an examination of the region
from many of the more accessible eminences
—from the Galenstock, the Grauhaupt, the Pitz
Languard, the Monte Confinale—or, better still,
from Mont Blanc, Monte Rosa, the Jungfrau, the
Finsteraarhorn, the Weisshorn, or the Matterhorn,
where local peculiarities are toned down, and the
operations of the powers which really made this
region what it is are alone brought into prominence
—must, I imagine, convince every physically-minded
man of the inability of any fracture theory to
account for the present conformation of the Alps.

A correct model of the mountains, with an un-
exaggerated vertical scale, produces the same effect
upon the mind as the prospect from one of he
highest peaks. We are apt to be influenced by
local phenomena which, though insignificant in
view of the general question of Alpine conformation,
are, with reference to our customary standards, vast

and impressive. In a true model those local peculiarities disappear; for on the scale of a model they
are too small to be visible; while the essential facts
and forms are presented to the undistracted attention.

A minute analysis of the phenomena strengthens
the conviction which the general aspect of the Alps
fixes in the mind. We find, for example, numerous
valleys which the most ardent plutonist would not
think of ascribing to any other agency than erosion.
That such is their genesis and history is as certain
as that erosion produced the Chines in the Isle of
Wight. From these indubitable cases of erosion—
commencing, if necessary, with the small ravines
which run down the flanks of the ridges, with their
little working navigators at their bottoms—we can
proceed, by almost insensible gradations, to the
largest valleys of the Alps; and it would perplex
the plutonist to fix upon the point at which
fracture begins to play a material part.

In ascending one of the larger valleys, we enter it
where it is wide and where the eminences are gentle
on either side. The flanking mountains become
higher and more abrupt as we ascend, and at length
we reach a place where the depth of the valley is a
maximum. Continuing our walk upwards, we find
ourselves flanked by gentler slopes, and finally
emerge from the valley and reach the summit of an

open col, or depression in the chain of mountains. This is the common character of the large valleys. Crossing the col, we descend along the opposite slope of the chain, and through the same series of appearances in the reverse order. If the valleys on both sides of the col were produced by fissures, what prevents the fissure from prolonging itself across the col? The case here cited is representative; and I am not acquainted with a single instance in the Alps where the chain has been cracked in the manner indicated. The cols are simply depressions; and in the case of many of them the unfissured rock can be traced from side to side.

The typical instance just sketched follows as a natural consequence from the theory of erosion. Before either ice or water can exert great power as an erosive agent, it must collect in sufficient mass. On the higher slopes and plateaus—in the region of cols—the power is not fully developed; but lower down tributaries unite, erosion is carried on with increased vigour, and the excavation gradually reaches a maximum. Lower still the elevations diminish and the slopes become more gentle; the cutting power gradually relaxes, until finally the eroding agent quits the mountains altogether, and the grand effects which it produced in the earlier portions of its course entirely disappear.

I have hitherto confined myself to the consideration

of the broad question of the erosion theory as compared with the fracture theory; and all that I have been able to observe and think with reference to the subject leads me to adopt the former. Under the term erosion I include the action of water, of ice, and of the atmosphere, including frost and rain. Water and ice, however, are the principal agents, and which of these two has produced the greatest effect it is perhaps impossible to say. Two years ago I wrote a brief note 'On the Conformation of the Alps,'[1] in which I ascribed the paramount influence to glaciers. The facts on which that opinion was founded are, I think, unassailable; but whether the conclusion then announced fairly follows from the facts is, I confess, an open question.

The arguments which have been thus far urged against the conclusion are not convincing. Indeed, the idea of glacier erosion appears so daring to some minds that its boldness alone is deemed its sufficient refutation. It is, however, to be remembered that a precisely similar position was taken up by many respectable people when the question of ancient glacier extension was first mooted. The idea was considered too hardy to be entertained; and the evidences of glacial action were sought to be explained by reference to almost any process rather than the true one. Let those who so wisely took the side of

[1] Phil. Mag. vol. xxiv. p. 169.

'boldness' in that discussion beware lest they place
themselves, with reference to the question of glacier
erosion, in the position formerly occupied by their
opponents.

Looking at the little glaciers of the present day—
mere pigmies as compared to the giants of the
glacial epoch—we find that from every one of them
issues a river more or less voluminous, charged with
the matter which the ice has rubbed from the rocks.
Where the rocks are of a soft character, the amount
of this finely pulverised matter suspended in the
water is very great.   The water, for example, of the
river which flows from Santa Catarina to Bormio
is thick with it.   The Rhine is charged with this
matter, and by it has so silted up the Lake of
Constance as to abolish it for a large fraction of its
length.   The Rhone is charged with it, and tens of
thousands of acres of cultivable land are formed by
it above the Lake of Geneva.

In the case of every glacier we have two agents
at work—the ice exerting a crushing force on
every point of its bed which bears its weight, and
either rasping this point into powder or tearing it
bodily from the rock to which it belongs; while
the water which everywhere circulates upon the bed
of the glacier continually washes the detritus away
and leaves the rock clean for further abrasion.
Confining the action of glaciers to the simple rubbing

away of the rocks, and allowing them sufficient time to act, it is not a matter of opinion, but a physical certainty, that they will scoop out valleys. But the glacier does more than abrade. Rocks are not homogeneous ; they are intersected by joints and places of weakness, which divide them into virtually detached masses. A glacier is undoubtedly competent to root such masses bodily away. Indeed the mere à *priori* consideration of the subject proves the competence of a glacier to deepen its bed. Taking the case of a glacier 1,000 feet deep (and some of the older ones were probably three times this depth), and allowing 40 feet of ice to an atmosphere, we find that on every square inch of its bed such a glacier presses with a weight of 375 lbs., and on every square yard of its bed with a weight of 486,000 lbs. With a *vertical* pressure of this amount the glacier is urged down its valley by the pressure from behind. We can hardly, I think, deny to such a tool a power of excavation.

Before concluding these remarks, I refreshed my memory by a second reading of the paper of Mr. John Ball, published in the 'Philosophical Magazine' for February 1863. Mr. Ball's great experience of the Alps naturally renders everything he writes regarding them interesting. But though I have attended to the suggestions contained in his paper, I am unable to see the cogency of his arguments.

R

An inspection of the map of Switzerland, with reference to the direction of its valleys, suggests to my mind no objection whatever to the theory of erosion.

The reperusal of his paper assured me that Mr. Ball had paid attention to the formation of ancient lakes. He deems their beds a prominent feature of Alpine valleys; and he considers the barriers which dammed them up, and which were not removed by the ancient glaciers, as ' a formidable difficulty in the way of Prof. Tyndall's bold hypothesis.' ' Looking at the operation as a whole,' writes Mr. Ball, ' it is to me quite inconceivable that a glacier should be competent to scoop out valleys a mile or more in depth, and yet be unable to remove the main inequalities from its own channel.'

To this I reply that a glacier *is* competent to remove such barriers, and they probably have been ground down in some cases thousands of feet. But being of more resisting material than the adjacent rock, they are not ground down to the level of that rock. Were its bed uniform in the first instance, the glacier would, in my opinion, *produce* the inequalities which Mr. Ball thinks it ought to remove. I have recently had the pleasure of examining some of these barriers in the company of Mr. Ball; and to me they represented nothing more than the natural accidents of the

locality.   It would, I think, be far more wonderful
to find the rocks of the Alps perfectly homogeneous,
than to find them exhibiting such variations of resistance to grinding down as are actually observed.

The question of lake-basins is now in competent hands, and on its merits I will offer no
opinion. But I cannot help remarking that the
dams referred to by Mr. Ball furnish a conclusive
reply to some of the arguments which have been
urged against Prof. Ramsay's theory.   These barriers
have been crossed by the ice, and many of them
present steeper gradients than Prof. Ramsay has
to cope with in order to get his ice out of his lake-
basins.   An inspection of the barriers shows that
they were incompetent to embay the ice : they are
scarred and fluted from bottom to top.   When it is
urged against Prof. Ramsay that a glacier cannot
drop into a hole 2,000 feet deep and get out again,
the distance ought to be stated over which these
2,000 feet have to be distributed.   A depression
2,000 feet deep, if only of sufficient length, would
constitute no material obstacle to the motion of a
great glacier.

The retardation of a glacier by its bed has
also been referred to as proving its impotence as
an erosive agent; but this very retardation is in
some measure an expression of the magnitude of the
erosive energy.   Either the bed must give way, or

the ice must slide over itself; and to make ice slide over itself requires great power. We get some idea of the crushing pressure which the moving glacier exercises against its bed from the fact that the resistance, and the effort to overcome it, are such as to make the upper layers of a glacier move bodily over the lower ones—a portion only of the total motion being due to the progress of the entire mass of the glacier down its valley.

The sudden bend in the valley of the Rhone at Martigny has also been regarded as conclusive evidence against the theory of erosion. 'Why,' it has been asked, 'did not the glacier of the Rhone go straight forward instead of making this awkward bend?' But if the valley be a crack, why did the crack make this bend? The crack, I submit, had at least as much reason to prolong itself in a straight line as the glacier had. A statement of Sir John Herschel with reference to another matter is perfectly applicable here: 'A crack once produced has a tendency to run—for this plain reason, that at its momentary limit, at the point at which it has just arrived, the divellent force on the molecules there situated is counteracted only by half of the cohesive force which acted when there was no crack, viz. the cohesion of the uncracked portion alone' ('Proc. Roy. Soc.' vol. xii. p. 678). To account, then, for the bend, the adherent of the fracture theory must assume the existence of some accident which turned the crack at right angles to itself; and he surely will permit the adherent of the erosion theory to make a similar assumption.

The influence of small accidents on the direction of rivers is beautifully illustrated in glacier streams, which are made to cut either straight or sinuous channels by causes apparently of the most trivial character. In his interesting paper ' On the Lakes of Switzerland,' M. Studer also refers to the bend of the Rhine at Sargans in proof that the river must there follow a pre-existing fissure. I made a special expedition to the place in 1864; and though I felt that M. Studer had good grounds for the selection of this spot, I was unable to arrive at his conclusion as to the necessity of a fissure.

Again, in the interesting volume recently published by the Swiss Alpine Club, M. Desor informs us that the Swiss naturalists who met last year at Samaden visited the end of the Morteratsch glacier, and there convinced themselves that a glacier had no tendency whatever to imbed itself in the soil. I scarcely think that the question of glacier erosion, as applied either to lakes or valleys, is to be disposed of so easily. Let me record here my experience of the Morteratsch glacier. I took with me in 1864 a theodolite to Pontresina, and while there had to congratulate myself on the invaluable aid of my friend Mr. Hirst, who in 1857 did such good service upon the Mer de Glace and its tributaries. We set out three lines across the Morteratsch glacier, one of which crossed the ice-stream near the well-known hut of the painter Georgei, while the two others were staked out, the one above the hut and the other below it. Calling the highest line A, the line which crossed the glacier at the hut B, and the lowest line C, the following are the mean hourly motions of the three lines, deduced from observations which extended over several days. On each line eleven stakes were fixed, which are designated by the figures 1, 2, 3, &c. in the Tables.

Morteratsch Glacier, Line A.

No. of Stake.	Hourly Motion.
1	0·35 inch.
2	0·49 „
3	0·53 „
4	0·54 „
5	0·56 „
6	0·54 „
7	0·52 „
8	0·49 „
9	0·40 „
10	0·29 „
11	0·20 „

As in all other measurements of this kind, the retarding influence of the sides of the glacier is manifest: the centre moves with the greatest velocity.

Morteratsch Glacier, Line B.

No. of Stake.					Hourly Motion.
1	.	.	.	.	. 0·05 inch.
2	.	.	.	.	. 0·14 ,,
3	.	.	.	.	. 0·24 ,,
4	.	.	.	.	. 0·32 ,,
5	.	.	.	.	. 0·41 ,,
6	.	.	.	.	. 0·44 ,,
7	.	.	.	.	. 0·44 ,,
8	.	.	.	.	. 0·45 ,,
9	.	.	.	.	. 0·43 ,,
10	.	.	.	.	. 0·44 ,,
11	.	.	.	.	. 0·44 ,,

The first stake of this line was quite close to the edge of the glacier, and the ice was thin at the place, hence its slow motion. Crevasses prevented us from carrying the line sufficiently far across to render the retardation of the further side of the glacier fully evident.

Morteratsch Glacier, Line C.

No. of Stake.					Hourly Motion.
1	.	.	.	.	. 0·05 inch.
2	.	.	.	.	. 0·09 ,,
3	.	.	.	.	. 0·18 ,,
4	.	.	.	.	. 0·20 ,,
5	.	.	.	.	. 0·25 ,,
6	.	.	.	.	. 0·27 ,,
7	.	.	.	.	. 0·27 ,,
8	.	.	.	.	. 0·30 ,,
9	.	.	.	.	. 0·21 ,,
10	.	.	.	.	. 0 20 ,,
11	.	.	.	.	. 0·16 ,,

Comparing the three lines together, it will be observed that the velocity diminishes as we descend the glacier. In 100 hours the maximum motion of the three lines respectively is as follows :

Maximum Motion in 100 *hours.*

Line A . . . . 56 inches
" B . . . . 45 "
C . . . . 30 "

This deportment explains an appearance which must strike every observer who looks upon the Morteratsch from the Piz Languard, or from the new Bernina Road. A medial moraine runs along the glacier, commencing as a narrow streak, but towards the end the moraine extending in width, until finally it quite covers the terminal portion of the glacier. The cause of this is revealed by the foregoing measurements, which prove that a stone on the moraine where it is crossed by the line A approaches a second stone on the moraine where it is crossed by the line C with a velocity of twenty-six inches per one hundred hours. The moraine is in a state of longitudinal compression. Its materials are more and more squeezed together, and they must consequently move laterally and render the moraine at the terminal portion of the glacier wider than above.

The motion of the Morteratsch glacier, then, diminishes as we descend. The maximum motion of the third line is thirty inches in one hundred hours, or seven inches a day—a very slow motion; and had we run a line nearer to the end of the glacier, the motion would have been slower still. At the end itself it is nearly insensible. Now I submit that this is not the place to seek for the scooping power of a glacier. The opinion appears to be prevalent that it is the snout of a glacier that must act the part of ploughshare; and it is certainly an erroneous opinion. The scooping power will exert itself most where the weight, and consequently (other things being equal) the motion, is greatest. A glacier's snout often *rests upon* matter which has been scooped from the glacier's bed higher up. I therefore do not think that the inspection of what the end of a glacier does or does not accomplish can decide this question.

The snout of a glacier is potent to remove anything against which it can fairly abut; and this power, notwithstanding the slowness of the motion, manifests itself at the end of the Morteratsch glacier. A hillock, bearing pine-trees, was in front of the glacier when Mr. Hirst and myself inspected its end; and this hillock is being bodily removed by the thrust of the ice. Several of the trees are overturned; and in a few years, if the glacier continues its reputed advance, the mound will certainly be ploughed away.

I will here record a few other measurements executed on the Rosegg glacier : the line was staked
out across the trunk formed by the junction of the
Rosegg proper with the Tschierva glacier, a short
distance below the rocky promontory called Agaliogs.

Rosegg Glacier.

No. of Stake.	Hourly Motion.
1	0·01 inch.
2	0·05 „
3	0·07 „
4	0·10 „
5	0·11 „
6	0·13 „
7	0·14 „
8	0·18 „
9	0·24 „
10	0·23 „
11	0·24 „

This is an extremely slowly moving glacier ; the
maximum hardly amounts to seven inches a day.
Crevasses prevented us from continuing the line
quite across the glacier.

To return to the question of Alpine conformation :
it stands, I think, thus : We have, in the first
place, great valleys, such as those of the Rhine and
the Rhone, which we might conveniently call valleys
of the first order.   The mountains which flank
these main valleys are also cut by lateral valleys
running into the main one, and which may be

called valleys of the second order. When these
latter are examined, smaller valleys are found
running into them, which may be called valleys
of the third order. Smaller ravines and depressions,
again, join the latter, which may be called valleys
of the fourth order, and so on until we reach streaks
and cuttings so minute as not to merit the name
of valleys at all. At the bottom of every valley
we have a stream, diminishing in magnitude as the
order of the valley ascends, carving the earth and
carrying its materials to lower levels. We find
that the larger valleys have been filled for untold
ages by glaciers of enormous dimensions, always
moving, grinding down and tearing away the rocks
over which they passed. We have, moreover, on
the plains at the feet of the mountains, and in enormous quantities, the very matter derived from the
sculpture of the mountains themselves.

The plains of Italy and Switzerland are cumbered
by the *débris* of the Alps. The lower, wider, and
more level valleys are also filled to unknown depths
with the materials derived from the higher ones.
In the vast quantities of moraine-matter which
cumber many even of the higher valleys we have
also suggestions as to the magnitude of the erosion
which has taken place. This moraine-matter, moreover, can only in small part have been derived from
the falling of rocks *upon* the ancient glacier ; it is

in great part derived from the grinding and the ploughing-out of the glacier itself. This accounts for the magnitude of many of the ancient moraines, which date from a period when almost all the mountains were covered with ice and snow, and when, consequently, the quantity of moraine-matter derived from the naked crests cannot have been considerable.

The erosion theory ascribes the formation of Alpine valleys to the agencies here briefly referred to. It invokes nothing but true causes. Its artificers are still there, though, it may be, in diminished strength; and if they are granted sufficient time, it is demonstrable that they are competent to produce the effects ascribed to them. And what does the fracture theory offer in comparison? From no possible application of this theory, pure and simple, can we obtain the slopes and forms of the mountains. Erosion must in the long run be invoked, and its power therefore conceded. The fracture theory infers from the disturbances of the Alps the existence of fissures; and this is a probable inference. But that they were of a magnitude sufficient to determine the conformation of the Alps, and that they followed, as the Alpine valleys do, the lines of natural drainage of the country, are assumptions which do not appear to me to be justified either by reason or by observation.

There is a grandeur in the secular integration of small effects implied by the theory of erosion almost superior to that involved in the idea of a cataclysm. Think of the ages which must have been consumed in the execution of this colossal sculpture. The question may, of course, be pushed to further limits. Think of the ages which the molten earth required for its consolidation. But these vaster epochs lack sublimity through our inability to grasp them. They bewilder us, but they fail to make a solemn impression. The genesis of the mountains comes more within the scope of the intellect, and the majesty of the operation is enhanced by our partial ability to conceive it. In the falling of a rock from a mountain-head, in the shoot of an avalanche, in the plunge of a cataract, we often see more impressive illustrations of the power of gravity than in the motions of the stars. When the intellect has to intervene, and calculation is necessary to the building up of the conception, the expansion of the feelings ceases to be proportional to the magnitude of the phenomena.

XXI.

SEARCH ON THE MATTERHORN : A PROJECT.

In July 1865 my excellent friend Hirst and myself
visited Glarus, intending, if circumstances favoured
us, to climb the Tödi. We had, however, some
difficulty with the guides, and therefore gave the
expedition up. Crossing the Klausen pass to Altdorf,
we ascended the Gotthardt Strasse to Wasen, and
went thence over the Susten pass to Gadmen, which
we reached late at night. We halted for a moment
at Stein, but the blossom of 1863[1] was no longer
there, and we did not tarry. On quitting Gadmen
next morning I was accosted by a guide, who asked
me whether I knew Professor Tyndall. ' He is
killed, sir,' said the man—' killed upon the Matterhorn.' I then listened to a somewhat detailed
account of my own destruction, and soon gathered
that, though the details were erroneous, something
serious if not shocking had occurred. At Imhof the
rumour became more consistent, and immediately
afterwards the Matterhorn catastrophe was in every

[1] Page 169.

mouth, and in all the newspapers. My friend and myself wandered on to Mürren, whence, after an ineffectual attempt to cross the Petersgrat, we went by Kandersteg and the Gemmi to Zermatt.

Of the four sufferers on the Matterhorn one remained behind. But expressed in terms either of mental torture or physical pain, the suffering in my opinion was *nil.* Excitement during the first moments left no room for terror, and immediate unconsciousness prevented pain. No death has probably less of agony in it than that caused by the shock of gravity on a mountain-side. *Expected,* it would be terrible; but unexpected, not. I had heard, however, of other griefs and sufferings consequent on the accident, and this prompted a desire on my part to find the remaining one and bring him down.

I had seen the road-makers at work between St. Nicholas and Zermatt, and was struck by the rapidity with which they pierced the rocks for blasting. One of these fellows could drive a hole a foot deep into hard granite in less than an hour. I was therefore determined to secure in aid of my project the services of a road-maker. None of the Zermatt guides would second me, but I found one of the Lochmatters of St. Nicholas willing to do so. Him I sent to Geneva to buy 3,000 feet of rope, which duly came on heavily laden mules to Zermatt.

Hammers and steel punches were prepared; a tent was put in order, and the whole was carried up to the chapel by the Schwarz See. But the weather would by no means smile upon the undertaking. I waited in Zermatt for twenty days, making excursions with pleasant friends, but they merely spanned the brief intervals which separated one rain-gush or thunderstorm from another. Bound by an engagement to my friend Professor De la Rive, of Geneva, where the Swiss naturalists had their annual assembly in 1865, I was forced to leave Zermatt. My notion was to climb to the point where the men slipped, and to fix there suitable irons in the rocks. By means of ropes attached to these I proposed to scour the mountain along the line of the glissade. There were peculiarities in the notion which need not now be dwelt upon, inasmuch as the weather rendered them all futile.

[I am not sure that the proposed search is practicable; it would certainly require unusually good weather for its execution.—April 1871.]

XXII.

THE TITLIS, FINSTERAARSCHLUCHT, PETERSGRAT, AND ITALIAN LAKES.

In the summer of 1866 I first went to Engsteln, one of the most charming spots in the Alps. It had at that time a double charm, for the handsome young widow who kept the inn supplemented by her kindness and attention within doors the pleasures extracted from the outer world. A man named Maurer, of Meyringen, was my guide for a time. We climbed the Titlis, going straight up it from the Joch Pass, in the track of a scampering chamois which showed us the way. The Titlis is a very noble mass—one of the few which, while moderate in height, bear a lordly weight of snow. The view from the summit is exceedingly fine, and on it I repeated with a hand spectroscope the observations of M. Janssen on the absorption-bands of aqueous vapour. On the day after this ascent I quitted Engsteln, being drawn towards the Wellhorn and Wetterhorn, both of which, as seen from Engsteln, came out with inexpressible nobleness. The upper

s

dome of heaven was of the deepest blue, while only the faintest lightening of the colour towards the horizon indicated the augmented thickness of the atmosphere in that direction. The sun was very hot, but there was a clear rivulet at hand, deepening here and there into pebbled pools, into which I plunged at intervals, causing my guide surprise if not anxiety; for he shared the common super-stition that plunging, when hot, into cold water is dangerous. The danger, and a very serious one it is, is to plunge into cold water when *cold*. The strongest alone can then bear immersion without damage.

This year I subjected the famous Finsteraarschlucht to a closer examination than ordinary. The earth-quake theory already adverted to was prevalent regarding it, and I wished to see whether any evidences existed of aqueous erosion. It will be remembered that the Schlucht or gorge is cut through a great barrier of limestone rock called the Kirchet, which throws itself across the valley of Hasli, about three-quarters of an hour's walk above Meyringen. The plain beyond the barrier, on which stands the hamlet of Imhof, is formed of the sedi-ment of a lake of which the Kirchet constituted the dam. This dam is now cut through for the passage of the Aar, forming one of the noblest gorges in Switzerland. Near the summit of the Kirchet is a house with a signboard inviting the traveller to visit the *Aarenschlucht*, a narrow lateral gorge which runs down to the very bottom of the principal one. The aspect of this smaller chasm from its bottom to its top proves to demonstration that water had in former ages worked there as a navigator. It is scooped, rounded, and polished, so as to render it palpable to the common eye that it is a gorge of erosion. But it was regarding the sides of the great chasm that I needed instruction, and from its edge I could see nothing to satisfy me. I therefore stripped and waded into the river until a point was reached which commanded an excellent view of both sides of the gorge. The water was cutting, but I was repaid. Below me on the left-hand side was a jutting cliff, which bore the thrust of the river and caused the Aar to swerve from its direct course. From top to bottom this cliff was polished, rounded, and scooped. There was no room for doubt. The river which now runs so deeply down had once been above. It has been the delver of its own channel through the barrier of the Kirchet.

I went on to Rosenlaui, proposing to climb the neighbouring mountains in succession. In fact I went to Switzerland in 1866 with a particular hunger for the heights. But the weather thickened before Rosenlaui was reached, and on the night following the morning of my departure from Engsteln

I lay upon my plaid under an impervious pine, and watched as wild a thunderstorm and as heavy a downpour of rain as I had ever seen. Most extraordinary was the flicker on cliffs and trees, and most tremendous was the detonation succeeding each discharge. The fine weather came thus to an end, and next day I gave up the Wetterhorn for the ignoble Faulhorn. Here the wind changed, the air became piercingly cold, and on the following morning heavy snow-drifts buttressed the doors, windows, and walls of the inn. We broke away, sinking at some places to the hips in snow. A descent of a thousand feet carried us from the bleakest winter into genial summer. My companion held on to the beaten track, while I sought a rougher and more direct one to the Scheinigeplatte, a resting-place which commands a noble view of the precipices of the Jungfrau. We were solitary visitors there, and I filled the evening with Miss Thackeray's 'Story of Elizabeth,' which some benevolent traveller had left at the hotel.

Thence we dropped down to Lauterbrunnen, went up the valley to the little inn at Trechslawinen, and crossed the Petersgrat the following day. The recent precipitation had cleared the heavens and reloaded the heights. It was, perhaps, the splendour of the weather and the purity of the snows, aided by the subjective effect due to contrast with a series

of most dismal days, that made me think the Petersgrat so noble a standpoint for a view of the mountains. The horizontal extent was vast, and the
grouping magnificent. The undoubted monarch of
this unparagoned scene was the Weisshorn, and this
may have rendered me partial in my judgment, for
men like to see what they love exalted. At Platten
we found shelter in the house of the curé. Next
day we crossed the Lotschsattel, and swept round by
the Aletsch glacier to the Æggischhorn.

Here I had the pleasure of meeting a very ardent
climber, who entertains peculiar notions regarding
guides. He deems them, and rightly so, very ex-
pensive, and he also feels pleasure in trying his own
powers. Very likely it is my habit of going alone
that causes me to sympathise with him. I would,
however, admonish him that he may go too far in
this direction, and probably his own experience has
by this time forestalled the admonition. Still, if
skill, strength, and self-reliance are things to be
cultivated in the Alps, they are, within certain
limits, best exercised and developed in the absence
of guides. And if the real climbers are ever to be
differentiated from the crowd who write and talk
about the mountains, it is only to be done by dis-
pensing with professional assistance. But no man
without natural aptitude and due training would be
justified in committing himself to ventures of this

kind, and it is an error to suppose that the necessary
knowledge can be obtained in one or two summers
in the Alps. Climbing is an art, and those who
wish to cultivate it on their own account ought to
give themselves sufficient previous practice in the
company of first-rate guides. Here, moreover, as
in every other sphere of human action, whether
intellectual or physical, as indeed among the guides
themselves, real eminence falls only to the lot of
few. Whatever be the amount of preparation, real
climbers must still remain select men.

From the Bel Alp, Mr. Girdlestone and I, without
any guide, made an attack upon the Aletschhorn.
We failed. The weather as we started was undecided, but we hoped the turn might be in our
favour. We first kept along the Alp, with the Jäggi
glacier to our right, then crossed its moraine, and
made the trunk glacier our highway until we
reached the point of confluence of its branches.
Here we turned to the right, the Aletschhorn, from
base to summit, coming into view. We reached
the true base of the mountain, and without halting
breasted its snow. But as we climbed the atmosphere thickened more and more. About the Nesthorn the horizon deepened to pitchy darkness, and
on the Aletschhorn itself hung a cloud, which we at
first hoped would melt before the strengthening sun,
but which instead of melting became denser. Now

and then an echoing rumble of the wind warned us
that we might expect rough handling above. We
persisted, however, and reached a considerable
height, unwilling to admit that the weather was
against us, until a more savage roar and a ruder
shake than ordinary caused us to halt, and look
more earnestly and anxiously into the darkening
atmosphere. We were forced to give in, and during
our descent the air was thick and dark with falling
snow. Holding on in the dimness to the medial
moraine, we managed to get down the glacier, and
to clear it at a practicable point, whence, guided
by the cliffs which flanked our right, and which
became visible only when we came almost into
contact with them, we hit the proper track to the
Bel Alp hotel.

Though my visits to the Alps had already numbered
thirteen, I had never gone so far southward as the
Italian lakes. The perfectly unmanageable weather
of July 1866 caused me to cross with Mr. Girdlestone
into Italy, in the hope that a respite of ten or
twelve days might improve the temper of the
mountains. We walked over the Simplon to the
village of the same name, and took thence the
diligence to Domo d'Ossola and Baveno. The atmospheric change was wonderful; and still the
clear air which we enjoyed below was the self-same
air that heaped clouds and snow upon the mountains.

It came across the heated plains of Lombardy, charged with moisture, but the moisture was in the transparent condition of true vapour, and hence invisible. Tilted by the mountains, the air rose, and as it expanded it became chilled, and as it became chilled it discharged its vapour as visible cloud, the globules of which swelled by coalescence into rain-drops on the mountain-flanks, or were frozen to snow upon the mountain-heads.

We halted on the margin of the Lago Maggiore. I could hear the lisping of the waters on the shingle far into the night. My window looked eastward, and through it could be seen the first warming of the sky at the approach of dawn. I rose, and watched the growth of colour all along the east. The mountains, from mere masses of darkness projected against the heavens, became empurpled. It was not as a mere wash of colour overspreading their surfaces. They blent with the atmosphere as if they were part and parcel of the general purple of the air. Nobody was stirring at the time, and the 'lap' of the lake upon its shore only increased the sense of silence.

> The holy hour was quiet as a nun
> Breathless with adoration.

In my subsequent experience of the Italian lakes I met with nothing which affected me so deeply as this morning scene on the Lago Maggiore.

From Baveno we crossed the lake to Luino, and
went thence to Lugano. At Belaggio, on the junction of the two branches of the Lake of Como, we
halted a couple of days. Como itself we reached in
a small sailing-boat, as a storm prevented the steamer
from taking us. There we saw the statue of Volta
—a prophet justly honoured in his own country.
From Como we went to Milan. A climber, of course,
could not forego the pleasure of looking at Monte
Rosa from the cathedral roof. The distribution of
the statues magnified the apparent vastness of the
pile ; still the impression made on me by this great
edifice was one of disappointment. Its front seemed
to illustrate an attempt to cover meanness of conception by profusion of adornment. The interior, however, notwithstanding the cheat of the ceiling, is
exceedingly grand.

From Milan we went to Orta, where we had a
plunge into the lake. We crossed it subsequently,
and walked on to Varallo : thence by Fobello over a
country of noble beauty to Ponte Grande in the Val
Ansasca. Thence again by Macugnaga, over the
deep snow of the Monte Moro, reaching Mattmark
in drenching rain. The temper of the northern
slopes did not appear to have improved during our
absence. We returned to the Bel Alp, fitful
triumphs of the sun causing us to hope that we might
still have fair play upon the Aletschhorn. But the day

after our arrival snow fell so heavily as to cover the pastures for 2,000 feet below the hotel. Partial famine among the herds was the consequence. They had eventually to be driven below the snow-line. Avalanches were not unfrequent on slopes which a day or two previously had been covered with grass and flowers. In this condition of things Mr. Milman, Mr. Girdlestone and I, climbed the Sparrenhorn, and found its heavy-laden Kamm almost as hard as that of Monte Rosa. Occupation out of doors was, however, insufficient to fill the mind, so I wound my plaid around my loins, and in my cold bedroom studied ' Mozley upon Miracles.'

XXIII.

ASCENT OF THE EIGER AND PASSAGE OF THE TRIFT.

GRINDELWALD was my first halting-place in the summer of 1867: I reached it, in company with a friend, on Sunday evening the 7th of July. The air of the glaciers and the excellent little dinners of the Adler rendered me rapidly fit for mountain-work. The first day we made an excursion along the lower glacier to the Kastenstein, crossing, in returning, the Strahleck branch of the glacier above the ice-fall, and coming down by the Zäsenberg. The second day was spent upon the upper glacier. The sunset covered the crest of the Eiger with indescribable glory that evening. It gave definition to a vague desire I had previously entertained to climb the mountain, and I forthwith arranged with excellent old Christian Michel, and with Peter Baumann, the preliminaries of the ascent.

At half-past one o'clock on the morning of the 11th we started from the Wengern Alp; no trace of cloud was visible in the heavens, which were sown broadcast with stars. Those low down twinkled with extraordinary vivacity, many of them flashing

lights of different colours.   When an opera-glass
was pointed to such a star, and shaken, the line of
light described by the image of the star resolved
itself into a string of richly coloured beads : rubies
and emeralds hung thus together on the same curve.
The dark intervals between the beads corresponded
to the moments of extinction of the star.   Over the
summit of the Wetterhorn the Pleiades hung like a
diadem, while at intervals a solitary meteor shot
across the sky.

We passed along the Alp, and then over the balled
snow and broken ice cast down a glacier which
fronted us.   Here the ascent began ; we passed from
snow to rock and from rock to snow by turns.   The
steepness for a time was moderate, the only thing
requiring caution being the thin crusts of ice upon
the rocks over which water had trickled the previous
day.   The east gradually brightened, the stars became paler and disappeared, and at length the crown
of the adjacent Jungfrau rose out of the twilight
into the rose of the sun.   The bloom crept gradually
downwards over the snows.   At length the whole
mountain-world partook of the colour.   It is not in
the night nor in the day—it is not in any statical
condition of the atmosphere—that the mountains
look most sublime.   It is during the few minutes
of transition from twilight to full day through the
splendours of the dawn.

Seven hours' climbing brought us to the higher slopes, which were for the most part ice, and required deep step-cutting. The whole duty of the climber on such slopes is to cut his steps properly, and to stand in them securely. At one period of my mountain life I looked lightly on the possibility of a slip, having full faith in the resources of him who accompanied me, and very little doubt of my own. Experience has qualified this faith in the power even of the best of climbers upon a steep ice-slope. A slip under such circumstances must not occur.

The Jungfrau began her cannonade very early, five avalanches having thundered down her precipices before eight o'clock in the morning. Bauman, being the youngest man, undertook the labour of step-cutting, which the hardness of the ice rendered severe. He was glad from time to time to escape to the snow-cornice which, unsupported save by its own tenacity, overhung the Grindelwald side of the mountain, checking himself at intervals by looking over the edge of the cornice, to assure himself that its strength was sufficient to bear our weight. A wilder precipice is hardly to be seen than this wall of the Eiger, viewed from the cornice at its top. It seems to drop sheer for eight thousand feet down to Grindelwald. When the cornice became unsafe, the guide retreated, and step-cutting recommenced. We reached the summit before nine

o'clock, and had from it an outlook over as glorious
a scene as this world perhaps affords.

On the following day I went down to Lauterbrunnen, and afterwards crossed the Petersgrat to
Platten, where, the door of the curé being closed
against travellers, we were forced into dirty quarters
in an adjacent house. From Platten, instead of going
as before over the Lötschsattel, we struck obliquely
across the ridge above the Nesthorn, and got down
upon the Jäggi glacier, making thus an exceedingly
fine excursion from Platten to the Bel Alp. Thence,
after a day's halt, I pushed on to Zermatt.

I have already mentioned Carrel, *the bersaglier*,
who accompanied Bennen and myself in our attempt
upon the Matterhorn in 1862, and who in 1865
reached the summit of the mountain. With him
I had been in correspondence for some time, and
from his letters an enthusiastic desire to be my
guide up the Matterhorn might be inferred. From
the Riffelberg I crossed the Theodule to Breuil,
where I saw Carrel. He had naturally and de-
servedly grown in his own estimation. But I
was discomfited by the form his self-consciousness
assumed. His demands were exorbitant, and he
also objected to the excellent company of Christian
Michel. In fact my friend Carrel was no longer
a reasonable man. I believe he afterwards felt

ashamed of himself, and sent his friends Bich and Meynet to speak to me while he kept aloof. But the weather was then too bad to permit of any definite arrangement being made.

I waited at the Riffel for twelve days, making small excursions here and there. But, though the weather was not so abominable as it had been in the previous year, the frequent snow-discharges on the Matterhorn kept it unassailable. In company with Mr. Crawfurd Grove, who had engaged Carrel as his guide, Michel being mine, I made the pass of the Trift from Zermatt to Zinal. I could understand and share the enthusiasm experienced by Mr. Hinchliff in crossing this truly noble pass. It is certainly one of the finest in the whole Alps. For that one day, moreover, the weather was magnificent. Next day we crossed to Evolena, going considerably astray, and thus converting a light day into a rather heavy one. From Evolena we purposed crossing the Col d'Érin back to Zermatt, but the weather would not let us. This excursion had been made with the view of allowing the Matterhorn a little time to arrange its temper; but the temper continued sulky, and at length wearied me out. We went round by the valley of the Rhone to Zermatt, and, finding matters worse than ever, both Mr. Grove and myself returned to Visp, intending to quit Switzerland altogether. Here he changed his mind and returned to Zermatt;

on the same day the weather changed also, and continued fine for a fortnight. He succeeded in getting with Carrel to the top of the Matterhorn, and I succeeded in joining the British Association at Dundee. A ramble in the Highlands, including a visit to the Parallel Roads of Glenroy, concluded my vacation in 1867.

XXIV.

THE MATTERHORN—THIRD AND LAST ASSAULT.

THE oil of life burnt rather low with me in 1868. Driven from London by Dr. Bence Jones, I reached the Giessbach hotel on the Lake of Brientz early in July. No pleasanter position could be found for an invalid. My friend Hirst was with me, and we made various little excursions in the neighbourhood. The most pleasant of these was to the Hinterburger See, a small and lonely lake high up among the hills, fringed on one side by pines, and overshadowed on the other by the massive limestone buttresses of the Hinterburg. It is an exceedingly lovely spot, but rarely visited. The Giessbach hotel is an admirably organised establishment. The table is served by Swiss girls in Swiss costume, fresh, handsome, and modest, well brought up, who come there, not as servants, but to learn the mysteries of housekeeping. And among her maidens moved like a little queen the daughter of the host—noiseless, but effectual in her rule and governance. I went to the Giessbach with a prejudice against the

T

illumination of the fall. The crowd of spectators may suggest the theatre, but the lighting up of the water is fine. I liked the colourless light best; it merely intensified the contrast revealed by ordinary daylight between the white foam of the cascades and the black surrounding pines.

From the Giessbach we went to Thun, and thence up the Simmenthal to Lenk. Over the sulphur spring a large hotel has been recently erected, and here we found a number of Swiss and Germans, who thought the waters did them good. In one large room the liquid gushes from a tap into a basin, diffusing through the place the odour of rotten eggs. The patients like this smell; indeed they regard its foulness as a measure of their benefit. The director of the establishment is intelligent and obliging, sparing no pains to meet the wishes and promote the comfort of his guests. We wandered while at Lenk to the summit of the Rawyl pass, visited the Siebenbrünnen, where the river Simmen bursts full-grown from the rocks, and we should have clambered up the Wildstrubel had the weather been tolerable. From Lenk we went to Gsteig, a finely situated hamlet, but not celebrated for the peace and comfort of its inn; and from Gsteig to the Diablerets hotel. While there I clambered up the Diablerets mountain, and was amazed at the extent of the snow-field upon its tabular top. The peaks, if they ever existed, have been shorn away, and miles of flat *névé*, unseen from below, overspread their section.

From the Diablerets we drove down to Aigle. The Traubenkur had not commenced, and there was therefore ample space for us at the excellent hotel. We were compelled to spend a night at Martigny. I heard the trumpet of its famous mosquito, but did not feel its attacks. The following night was more pleasantly spent on the cool col of the Great St. Bernard. On Tuesday, July 21, we reached Aosta, and, in accordance with previous telegraphic arrangement, met there the Chanoine Carrel. Jean-Jacques Carrel, the old companion of Mr. Hawkins and myself, and others at Breuil, were dissatisfied with the behaviour of the *bersaglier* last year, and this feeling the Chanoine shared. He had written to me during the winter, stating that two new men had scaled the Matterhorn, and that they were ready to accompany me anywhere. He now drove, with Hirst and myself, to Chatillon, where at the noisy and comfortless inn we spent the night. Here Hirst quitted me, and I turned with the Chanoine up the valley to Breuil.

At Val Tournanche I saw a maiden niece of the Chanoine who had gone high up the Matterhorn, and who, had the wind not assailed her petticoats too roughly, might, it was said, have reached the top.

I can believe it. Her wrist was like a weaver's beam, and her frame seemed a mass of potential energy. The Chanoine had recommended to me as guides the brothers Joseph and Pierre Maquignaz, of Val Tournanche, his praises of Joseph as a man of unshaken coolness, courage, and capacity as a climber being particularly strong. Previous to reaching Breuil, I saw this Joseph, who seemed to divine by instinct my name and aim.

Carrel was at Breuil, looking very dark; Bich petitioned for a porter's post, blaming Carrel bitterly for his greed in the previous year; but I left the arrangement of these matters wholly in the hands of Maquignaz. He joined me in the evening, and on the following day we ascended one of the neighbouring summits, discussing as we went our chances on the Matterhorn. In 1867 the chief precipitation took place in a low atmospheric layer, the base of the mountain being heavily laden with snow, while the summit and the higher rocks were bare. In 1868 the distribution was inverted, the top being heavily laden and the lower rocks clear. An additional element of uncertainty was thus introduced. Maquignaz could not say what obstacles the snow might oppose to us above, but he was resolute and hopeful. My desire was to finish for ever my contest with the Matterhorn by making a pass over its summit from Breuil to Zermatt. In this attempt my guide expressed his willingness to join me, his interest in the project being apparently equal to my own.

He, however, only knew the Zermatt side of the mountain through inspection from below; and he acknowledged that a dread of it had filled him the previous year. He now reasoned, however, that as Mr. Whymper and the Taugwalds had managed to descend, we ought to be able to do the same. On the Friday we climbed to the Col de la Furka, examined from it the northern face of the pyramid, and discovered the men who were engaged in building the cabin on that side. We worked afterwards along the ridge which stretches from the Matterhorn to the Theodule, crossing its gulleys and scaling all its heights. It was a pleasant piece of discipline, on new ground, to both my guide and me.

On the Thursday evening a violent thunderstorm had burst over Breuil, discharging new snow upon the heights, but also clearing the oppressive air. Though the heavens seemed clear in the early part of Friday, clouds showed a disposition to meet us from the south as we returned from the col. I enquired of my companion whether, in the event of the day being fine, he would be ready to start on Sunday. His answer was a prompt negative. In Val Tournanche, he said, they always 'sanctified the Sunday.' I mentioned Bennen, my pious Catholic guide, whom I permitted and encouraged to attend his mass on all possible occasions, but who, never-theless, always yielded without a murmur to the demands of the weather. The reasoning had its effect. On Saturday Maquignaz saw his confessor, and arranged with him to have a mass at 2 A.M. on Sunday; after which, unshaded by the sense of duties unperformed, he would commence the ascent.

The claims of religion being thus met, the point of next importance, that of money, was set at rest by my immediate acceptance of the tariff published by the Chanoine Carrel. The problem being thus reduced to one of muscular physics, we pondered the question of provisions, decided on a bill of fare, and committed its execution to the industrious mistress of the hotel.

A fog, impenetrable to vision, had filled the whole of the Val Tournanche on Saturday night, and the mountains were half concealed and half revealed by this fog when we rose on Sunday morning. The east at sunrise was louring, and the light which streamed through the cloud orifices was drawn in ominous red bars across the necks of the mountains. It was one of those uncomfortable Laodicean days which engender indecision—threatening, but not sufficiently so to warrant postponement. Two guides and two porters were considered necessary for the first day's climb. A volunteer, moreover, attached himself to our party, who carried a sheepskin as part

of the furniture of the cabin. To lighten their
labour, the porters took a mule with them as far as
the quadruped could climb, and afterwards divided
the load among themselves. While they did so I
observed the weather. The sun had risen with considerable power, and had broken the cloud-plane
to pieces. The severed clouds gathered into masses
more or less spherical, and were rolled grandly over
the ridges into Switzerland. Save for a swathe of
fog which now and then wrapped its flanks, the
Matterhorn itself remained clear, and strong hopes
were raised that the progress of the weather was in
the right direction.

We halted at the base of the Tête du Lion, a bold
precipice formed by the sudden cutting down of the
ridge which flanks the Val Tournanche to the right.
From its base to the Matterhorn stretches the Col
du Lion, crossed for the first time in 1860, by Mr.
Hawkins, myself, and our two guides. We were now
beside a snow-gulley, which was cut by a deep furrow
along its centre, and otherwise scarred by the descent
of stones. Here each man arranged his bundle and
himself so as to cross the gulley in the minimum of
time. The passage was safely made, a few flying
shingle only coming down upon us. But danger
declared itself where it was not expected. Joseph
Maquignaz led the way up the rocks. I was next,
Pierre Maquignaz next, and last of all the porters.

Suddenly a yell issued from the leader : ' *Cachez-vous !* ' I crouched instinctively against the rock, which formed a by no means perfect shelter, when a boulder buzzed past me through the air, smote the rocks below me, and with a savage hum flew down to the lower glacier.   Thus warned, we swerved to an *arête*, and when stones fell afterwards they plunged to the right or left of us.

In 1860 the great couloir which stretches from the Col du Lion downwards was filled with a *névé* of deep snow.   But the atmospheric conditions which have caused the glaciers of Switzerland to shrink so remarkably during the last ten years [1] have swept away this *névé*. We had descended it in 1860 hip-deep in snow, and I was now reminded of its steepness by the inclination of its bed.   Maquignaz was incredulous when I pointed out to him the line of descent to which we had been committed, in order to avoid the falling stones of the Tête du Lion. Bennen's warnings on the occasion were very emphatic, and I could understand their wisdom now better than I did them.

When Mr. Hawkins and myself first tried the

---

[1] I should estimate the level of the Lower Grindelwald glacier, at the point where it is usually entered upon to reach the Eismeer, to be nearly one hundred feet vertically lower in 1867 than it was in 1856.   I am glad to find that the question of ' Benchmarks ' to fix such changes of level is now before the Council of the British Association.

Matterhorn, a temporary danger, sufficient to quell for a time the enthusiasm even of our lion-hearted guide, was added to the permanent ones. Fresh snow had fallen two days before; it had quite over-sprinkled the Matterhorn, converting the brown of its crags into an iron-grey; this snow had been melted and refrozen, forming upon the rocks an enamelling of ice. Besides their physical front, moreover, in 1860, the rocks presented a psychological one, derived from the rumour of their savage inaccessibility. The crags, the ice, and the character of the mountain, all conspired to stir the feelings. Much of the wild mystery has now vanished, especially at those points which in 1860 were places of virgin difficulty, but down which ropes now hang to assist the climber. The intrinsic grandeur of the Matterhorn, however, cannot be effaced.

After some hours of steady climbing we halted upon a platform beside the tattered remnant of one of the tents employed by me in 1862. Here we sunned ourselves for an hour. We subsequently worked upward, scaling the crags and rounding the bases of those wild and wonderful rock-towers, into which the weather of ages has hewn the southern ridge of the Matterhorn. The work required knowledge, but with a fair amount of skill it is safe work. I can fancy nothing more fascinating to a man given by nature and habit to such things than a climb

alone among these crags and precipices. He need
not be *theological*, but, if complete, the grandeur of
the place would certainly fill him with religious awe.

Looked at from Breuil, the Matterhorn presents
two summits—the one, the summit proper, a square
rock-tower in appearance; the other, which is really
the end of a sharp ridge abutting against the rock-
tower, an apparently conical peak. On this peak
Bennen and myself planted our flagstaff in 1862.
At some distance below it the mountain is crossed by
an almost horizontal ledge, always loaded with snow,
which, from its resemblance to a white necktie, has
been called the *Cravate*. On this ledge a cabin was
put together in 1867. It stands above the precipice
where I quitted my rope in 1862. Up this precipice,
by the aid of a thicker—I will not say a stronger—
rope, we now scrambled, and, following the exact
route pursued by Bennen and myself five years
previously, we came to the end of the Cravate. At
some places the snow upon the ledge fell steeply
from its junction with the cliff; deep step-cutting
was also needed where the substance had been melted
and recongealed. The passage, however, was soon
accomplished along the Cravate to the cabin, which
was almost filled with snow.

Our first need was water. We could, of course,
always melt the snow, but this would involve a
wasteful expenditure of heat. The cliff at the base

of which the hut was built, overhung, and from its
edge the liquefied snow fell in showers beyond the
cabin.   Four ice-axes were fixed on the ledge, and
over them was spread the residue of a second tent
which I had left at Breuil in 1862.   The water
falling upon the canvas flowed towards its centre.
Here an orifice was made, through which the
liquid descended into vessels placed to receive it.
Some modification of this plan might probably be
employed with profit for the storing-up of water for
droughty years in England.

I lay for some hours in the warm sunshine, in
presence of the Italian mountains, watching the
mutations of the air.   But when the sun sank the
air became chill, and we all retired to the cabin.
We had no fire, though warmth was much needed.
A lover of the mountains, and of his kind, had
contributed an india-rubber mattrass, on which I
lay down, a light blanket being thrown over me,
while the guides and porters were rolled up in
sheepskins.   The mattress was a poor defence against
the cold of the subjacent rock.   I bore this for two
hours, unwilling to disturb the guides, but at
length it became intolerable.   On learning my
condition, however, the good fellows were soon
alert, and, folding a sheepskin round me, restored
me gradually to a pleasant temperature.   I fell
asleep, and found the guides preparing breakfast,

and the morning well advanced, when I opened my eyes.

It was past six o'clock when the two brothers and I quitted the cabin. The porters deemed their work accomplished, but they halted for a time to ascertain whether we were likely to be driven back or to push forward. We skirted the Cravate, and reached the ridge at its western extremity. This we ascended along the old route of Bennen and myself to the conical peak already referred to, which, as seen from Breuil, constitutes a kind of second summit of the Matterhorn. From this point to the base of the final precipice of the mountain stretches an *arête*, terribly hacked by the weather, but on the whole horizontal. When I first made the acquaintance of this savage ridge—called by Italians the Spalla—it was almost clear of snow. It was now loaded, the snow being bevelled to an edge of exceeding sharpness. The slope to the left, falling towards Zmutt, was exceedingly steep, while the precipices on the right were abysmal. No other part of the Matterhorn do I remember with greater interest than this. It was terrible, but its difficulties were fairly within the grasp of human skill, and this association is more ennobling than where the circumstances are such as to make you conscious of your own helplessness. On one of the sharpest teeth of the ridge Joseph Maquignaz

halted, and, turning to me with a smile, remarked,
'There is no room for giddiness here, sir.' In fact,
such possibilities, in such places, must be altogether excluded from the chapter of accidents of the
climber.

It was at the end of this ridge, where it abuts
against the last precipice of the Matterhorn, that
my second flagstaff was left in 1862. I think there
must have been something in the light falling upon
this precipice that gave it an aspect of greater
verticality when I first saw it than it seemed to
possess on the present occasion. We had, however,
been struggling for many hours previously, and may
have been dazed by our exertion. I cannot otherwise account for three of my party declining flatly
to make any attempt upon the precipice. It looks
very bad, but no real climber with his strength
unimpaired would pronounce it, without trial, insuperable. Fears of this rock-wall, however, had been
excited long before we reached it. It was probably
the addition of the psychological element to the
physical—the reluctance to encounter new dangers
on a mountain which had hitherto inspired a super-
stitious fear—that quelled further exertion.

Seven hundred feet, if the barometic measurement
can be trusted, of very difficult rock-work now lay
above us. In 1862 this height had been under-
estimated by both Bennen and myself. Of the

14,800 feet of the Matterhorn, we then thought we had accomplished 14,600. If the barometer speaks truly, we had only cleared 14,200.

Descending the end of the ridge, we crossed a narrow cleft, and grappled with the rocks at the other side of it. Our ascent was oblique, bearing to the right. The obliquity at one place fell to horizontality, and we had to work on the level round a difficult protuberance of rock. We cleared the difficulty without haste, and then rose straight against the precipice. Above us a rope hung down the cliff, left there by Maquignaz on the occasion of his first ascent. We reached the end of this rope, and some time was lost by my guide in assuring himself that it was not too much frayed by friction. Care in testing it was doubly necessary, for the rocks, bad in themselves, were here crusted with ice. The rope was in some places a mere hempen core surrounded by a casing of ice, over which the hands slid helplessly. Even with the aid of the rope in this condition it required an effort to get to the top of the precipice, and we willingly halted there to take a minute's breath. The ascent was virtually accomplished, and a few minutes more of rapid climbing placed us on the lightning-smitten top. Thus ended the long contest between me and the Matterhorn.

The day thus far had swung through alternations of fog and sunshine. While we were on the ridge

below, the air at times was blank and chill with
mist; then with rapid solution the cloud would
vanish, and open up the abysses right and left of us.
On our attaining the summit a fog from Italy
rolled over us, and for some minutes we were clasped
by a cold and clammy atmosphere. But this passed
rapidly away, leaving above us a blue heaven, and
far below us the sunny meadows of Zermatt. The
mountains were almost wholly unclouded, and such
clouds as lingered amongst them only added to
their magnificence. The Dent d'Érin, the Dent
Blanche, the Gabelhorn, the Mischabel, the range
of heights between it and Monte Rosa, the Lyskamm,
and the Breithorn, were all at hand, and clear;
while the Weisshorn, noblest and most beautiful of
all, shook out a banner towards the north, formed
by the humid southern air as it grazed the crest of
the mountain.

The world of peaks and glaciers surrounding this
immediate circlet of giants was also open to us up
to the horizon. Our glance over it was brief, for it
was eleven o'clock, and the work before us soon
claimed all our attention. I found the *débris* of
my former expedition everywhere — below, the
fragments of my tents, and on the top a piece of
my ladder fixed in the snow as a flagstaff. The
summit of the Matterhorn is a sharp horizontal
arête, and along this we now moved eastward. On

our left was the roof-like slope of snow seen from
the Riffel and Zermatt; on our right were the
savage precipices which fall into Italy. Looking to
the further end of the ridge, the snow there seemed
to be trodden down, and I drew my companions'
attention to the apparent footmarks. As we approached the place it became evident that human
feet had been there two or three days previously.
I think it was Mr. Elliot of Brighton[1] who had
made this ascent—the first accomplished from
Zermatt since 1865. On the eastern end of the
ridge we halted to take a little food—not that I
seemed to need it: it was the remonstrance of
reason rather than the consciousness of physical
want that caused me to do so.

We took our ounce of nutriment and gulp of
wine (my only sustenance during the entire day),
and stood for a moment silently and earnestly
looking down towards Zermatt. There was a certain official formality in the manner in which the
guides turned to me and asked, 'Êtes-vous content
d'essayer?' A sharp responsive 'Oui!' set us immediately in motion. It was nearly half-past eleven
when we quitted the summit. The descent of the
roof-like slope already referred to offered no difficulty; but the gradient very soon became more
formidable.

[1] Killed in 1869 upon the Schreckhorn.

One of the two faces of the Matterhorn pyramid, seen from Zermatt, falls towards the Zmutt glacier, and has a well-known snow-plateau at its base. The other face falls towards the Furgge glacier. We were on the former. For some time, however, we kept close to the *arête* formed by the intersection of the two faces of the pyramid, because nodules of rock jutted from it which offered a kind of footing. These rock protuberances helped us in another way : round them an extra rope which we carried was frequently doubled, and we let ourselves down by the rope as far as it could reach, liberating it afterwards (sometimes with difficulty) by a succession of jerks. In the choice and use of these protuberances the guides showed both judgment and skill. The rocks became gradually larger and more precipitous, a good deal of time being consumed in dropping down and doubling round them. Still we preferred them to the snow-slope at our left as long as they continued practicable.

This they at length ceased to be, and we had to commit ourselves to the slope. It was in the worst possible condition. When snow first falls at these great heights it is usually dry, and has no coherence. It resembles, to some extent, flour, or sand, or saw-dust. Shone upon by a strong sun it partly melts, shrinks, and becomes more consolidated, and when subsequently frozen it may be safely trusted. Even though the melting of the snow and its subsequent

U

freezing may be only very partial, the cementing of
the granules adds immensely to the safety of the
footing.   Hence the advantage of descending such a
slope before the sun has had time to unlock the
rigidity of the night's frost.   But we were on the
steepest Matterhorn slope during the two hottest
hours of the day, and the sun had done his work
effectually.   The layer of snow was about fifteen
inches thick.   In treading it we came immediately
upon the rock, which in most cases was too
smooth to furnish either prop or purchase.   It was
on this slope that the Matterhorn catastrophe occurred: it is on this slope that other catastrophes
will occur, if this mountain should ever become
fashionable.

Joseph Maquignaz was the leader of our little
party, and a brave, cool, and competent leader he
proved himself to be.   He was silent, save when he
answered his brother's anxious and oft-repeated
question, 'Es-tu bien placé, Joseph?'   Along with
being perfectly cool and brave, he seemed to be
perfectly truthful.   He did not pretend to be 'bien
placé' when he was not, nor avow a power of holding which he knew he did not possess.   Pierre
Maquignaz is, I believe, under ordinary circumstances, an excellent guide, and he enjoys the
reputation of being never tired.   But in such circumstances as we encountered on the Matterhorn he

is not the equal of his brother. Joseph, if I may use the term, is a man of high boiling point, his constitutional *sangfroid* resisting the ebullition of fear. Pierre, on the contrary, shows a strong tendency to boil over in perilous places.

Our progress was exceedingly slow, but it was steady and continued. At every step our leader trod the snow cautiously, seeking some rugosity on the rock beneath it. This, however, was rarely found, and in most cases he had to establish a mechanical attachment between the snow and the slope which bore it. No semblance of a slip occurred in the case of any one of us, and had it occurred I do not think the worst consequences could have been avoided. I wish to stamp this slope of the Matterhorn with the character that really belonged to it when I descended it, and I do not hesitate to say that the giving way of any one of our party would have carried the whole of us to ruin. Why, then, it may be asked employ the rope ? The rope, I reply, notwithstanding all its possible drawbacks under such circumstances, is the safeguard of the climber. Not to speak of the moral effect of its presence, an amount of help upon a dangerous slope that might be measured by the gravity of a few pounds is often of incalculable importance ; and thus, though the rope may be not only useless but disastrous if the footing be clearly lost, and the glissade fairly begun, it lessens immensely the chance of this occurrence.

With steady perseverance, difficulties upon a mountain, as elsewhere, come to an end. We were finally able to pass from the face of the pyramid to its rugged edge, where it was a great relief to feel that honest strength and fair skill, which might have gone for little on the slope, were masters of the situation.

Standing on the *arête*, at the foot of a remarkable cliff-gable seen from Zermatt, and permitting the vision to range over the Matterhorn, its appearance is exceedingly wild and impressive. Hardly two things can be more different than the two aspects of the mountain from above and below. Seen from the Riffel, or Zermatt, it presents itself as a compact pyramid, smooth and steep, and defiant of the weathering air. From above, it seems torn to pieces by the frosts of ages, while its vast facettes are so foreshortened as to stretch out into the distance like plains. But this under-estimate of the steepness of the mountain is checked by the deportment of its stones. Their discharge along the side of the pyramid to-day was incessant, and at any moment, by detaching a single boulder, we could let loose a cataract of them, which flew with wild rapidity and with a thunderous clatter down the mountain. We once wandered too far from the *arête*, and were

warned back to it by a train of these missiles sweeping past us.

As long as our planet yields less heat to space
than she receives from the bodies of space, so long
will the forms upon her surface undergo mutation ;
and as soon as equilibruim, in regard to heat, has
been established, we shall have, as Thomson has
pointed out, not peace, but death. Life is the product and accompaniment of change, and the self-
same power that tears the flanks of the hills to pieces
is the mainspring of the animal and vegetable worlds.
Still, there is something chilling in the contemplation of the irresistible and remorseless character of
those infinitesimal forces, whose integration through
the ages pulls down even the Matterhorn. Hacked
and hurt by time, the aspect of the mountain from
its higher crags saddened me. Hitherto the impression that it made was that of savage strength, but
here we had inexorable decay.

This notion of decay, however, implied a reference
to a period when the Matterhorn was in the full
strength of mountainhood. My thoughts naturally
ran back to its possible growth and origin. Nor
did they halt there, but wandered on through molten
worlds to that nebulous haze which philosophers have
regarded, and with good reason, as the proximate
source of all material things. I tried to look at
this universal cloud, containing within itself the

prediction of all that has since occurred; I tried to imagine it as the seat of those forces whose action was to issue in solar and stellar systems, and all that they involve. Did that formless fog contain potentially the *sadness* with which I regarded the Matterhorn? Did the *thought* which now ran back to it simply return to its primeval home? If so, had we not better recast our definitions of matter and force? for if life and thought be the very flower of both, any definition which omits life and thought must be inadequate, if not untrue.

Questions like these, useless as they seem, may still have a practical outcome. For if the final goal of man has not been yet attained, if his development has not been yet arrested, who can say that such yearnings and questionings are not necessary to the opening of a finer vision, to the budding and the growth of diviner powers? Without this upward force could man have risen to his present height? When I look at the heavens and the earth, at my own body, at my strength and weakness of mind, even at these ponderings, and ask myself, Is there no being or thing in the universe that knows more about these matters than I do?—what is my answer? Supposing our theologic schemes of creation, condemnation, and redemption to be dissipated; and the warmth of denial which they excite, and which, as a motive force, can match the warmth of

affirmation, dissipated at the same time ; would the
undeflected human mind return to the meridian of
absolute neutrality as regards these ultra-physical
questions ?   Is such a position one of stable equilibrium ?   Such are the questions, without replies,
which could run through consciousness during a
ten minutes' halt upon the weathered spire of the
Matterhorn.

We shook the rope away from us, and went
rapidly down the rocks.   The day was well advanced
when we reached the cabin, and between it and the
base of the pyramid we missed our way.   It was late
when we regained it, and by the time we reached the
ridge of the Hörnli we were unable to distinguish
rock from ice.   We should have fared better than
we did if we had kept along the ridge and felt our
way to the Schwarz See, whence there would have
been no difficulty in reaching Zermatt, but we left
the Hörnli to our right, and found ourselves incessantly checked in the darkness by ledges and precipices, possible and actual.   We were afterwards
entangled in the woods of Zmutt, carving our way
wearily through bush and bramble, and creeping
at times along dry and precipitous stream-beds.
But we finally struck the path and followed it to
Zermatt, which we reached between one and two
o'clock in the morning.

Having work to do for the Norwich meeting of

the British Association, I remained several days at the Riffel, taking occasional breathings with pleasant companions upon the Riffelhorn. I subsequently crossed the Weissthor with Mr. Paris to Mattmark, and immediately afterwards returned to England.

On the 4th of September, Signor Giordano, to whom we are indebted for a very complete geological section of the Matterhorn, with Joseph Maquignaz and Carrel as guides, followed my route over the mountain. In a letter dated Florence, December 31, 1868, he writes to me thus:

'Quant à moi, je dirai que vraiment, j'ai trouvé cette fois le pic assez difficile. . . . J'ai surtout trouvé difficile la traversée de l'arête qui suit le pic Tyndall du côté de l'Italie. Quant au versant suisse, je l'ai trouvé moins difficile que je ne croyais, parce que la neige y était un peu consolidée par la chaleur. En descendant le pic du côté de Zermatt j'ai encouru un véritable danger par les avalanches de pierres. . . . . Un de mes deux guides a eu le havresac coupé en deux par un bloc, et moi-même j'ai été un peu contusionné.'

XXV.

ASCENT OF THE ALETSCHHORN.

THE failure through bad weather of a former attempt upon the Aletschhorn has been already recorded ; but a succession of cloudless days at the Bel Alp in August 1869 stirred up the desire to try again. This was strengthened by the wish to make a series of observations from the greatest accessible elevation on the colour and polarisation of the sky. I had no guide of my own, but the Knecht at the hotel had been up the mountain, and I thought that we two might accomplish the ascent without any other assistance. It was the first time the mountain had been attempted by a single guide, and I was therefore careful to learn whether he was embarrassed by either doubt or fear. There was no doubt or fear in the matter : he really wished to go with me. His master (the proprietor of the hotel) had asked him whether he was not undertaking too much. ' I am undertaking no more than my companion,' was his reply.

At twenty minutes past two we quitted the Bel

Alp. The moon, which seven hours previously had cleared the eastern mountain-tops with a visible motion, was now sloping to the west. The light was white and brilliant, and shadows of corresponding darkness were cast upon the earth. The larger stars were out, those near the horizon especially sparkling with many-coloured fires. The Pleiades were near the zenith, while Orion hung his sword a few degrees above the eastern horizon. Our path lay along the slope of the mountain, parallel to the Oberaletsch glacier, the lateral moraine of which was close to us on our right. After climbing sundry grass acclivities we mounted this moraine, and made it our pathway for a time. At a certain point the shingly ridge became depressed, opening a natural passage to the glacier. We found the ice 'hummocky,' and therefore crossed it to a medial moraine composed of granite *débris* and loaded here and there with clean granite blocks of enormous size. Beyond this moraine we found smoother ice and better light, for we had previously journeyed in the shadow of the mountains.

We marched upwards along the glacier chatting sociably at times, but at times stilled into silence by the stillness of the night. 'Es tagt!' at length exclaimed my companion. It dawns! Orion had moved upwards, leaving space between him and the horizon for the morning star. All the east was

belted by that ' daffodil sky' which in some states
of the atmosphere announces the approach of day in
the Alps.   We spun towards the east.   It brightened
and deepened, but deeper than the orange of the
spectrum it did not fall.   Amid this the mountains
rose.   Silently and solemnly their dark and dented
outlines rested against the dawn.

The mass of light thus thrown over the shaded
earth long before the sun appeared above the horizon
came not from illuminated *clouds*, but from matter
far more attenuated than clouds—matter which maintains comparative permanence in the atmosphere,
while clouds are formed and dissipated.   It is not
light reflected from concentric shells of air of varying
density, of which our atmosphere may be rightly
assumed to be made up ; for the light reflected from
these convex layers is thrown, not upon the earth at
all, but into space.   The 'rose of dawn' is usually
ascribed, and with sufficient correctness, to *transmitted* light, the blue of the sky to *reflected* light ;
but in each case there is both transmission and
reflection.   No doubt the daffodil and orange of
the east this morning must have been transmitted
through long reaches of atmospheric air, and no
doubt it was during this passage of the rays that the
elective winnowing of the light occurred which gave
the sky its tint and splendour.   But if the distance
of the sun below the horizon when the dawn first

appeared be taken into account, it will become evident that the solar rays must have been caused to
swerve from their rectilineal course by *reflection*.
The *refraction* of the atmosphere would be wholly
incompetent to bend the rays round the convex
earth to the extent now under contemplation.

Thus the light which is reflected must be first
transmitted to the reflecting particles, while the
transmitted light, except in the direct line of the
sun, must be reflected to reach the eyes. What
mainly holds the light in our atmosphere after the
sun has retired behind the earth is, I imagine, the
suspended matter which produces the blue of the
sky and the morning and the evening red. Through
the reverberation of the rays from particle to particle,
there must be at the very noon of night a certain
amount of illumination. Twilight must continue
with varying degrees of intensity all night long, and
the visibility of the nocturnal firmament itself may
be due, not, as my excellent friend Dove seems
to assume, to the light of the stars, but in great part
to the light of the sun, scattered in all directions
through the atmosphere by the almost infinitely
attenuated matter held there in suspension.

We had every prospect of a glorious day. To our
left was the almost full moon, now close to the
ridge of the Sparrenhorn. The firmament was as
blue as ever I have seen it—deep and dark, and to

all appearance *pure*; that is to say, unmixed with
any colour of a lower grade of refrangibility than
the blue. The lunar shadows had already become
weak, and were finally washed away by the light of
the east. But while the shadows were at their
greatest depth, and therefore least invaded by the
dawn, I examined the firmament with a Nicol's prism.[1]
The moonlight, as I have said, came from the left,
and right in front of me was a mountain of dark
brown rock, behind which spread a heaven of the
most impressive depth and purity. I looked over
the mountain-crest through the prism. In one
position of the instrument the blue was not sensibly
affected ; in the rectangular position it was so far
quenched as to reduce the sky and the dark mountain beneath it to the same uniform hue. The
outline of the mountain could hardly be detached
from the sky above it. This was the direction in
which the prism showed its maximum quenching
power ; in no other direction was the extinction of
the light of the sky so perfect. And it was at right
angles to the lunar rays: so that, as regards the
polarisation of the sky, the beams of the moon
behave exactly like those of the sun.

The glacier along which we first marched was a
trunk of many tributaries, and consequently of many
' medial moraines,' such moraines being always *one*

---

[1] Art. X. of 'Fragments of Science' is devoted to the sky.

less in number than the tributaries.[1] But two
principal branches absorbed all the others as con-
stituents.   One of these descended from the Great
and Little Nesthorn and their spurs; the other
from the Aletschhorn. Up this latter branch we
steered from the junction. Hitherto the surface of
the glacier, disintegrated by the previous day's sun,
and again hardened by the night's frost, had crackled
under our feet; but on the Aletschhorn branch the
ice was coated by a kind of fur, resembling the nap
of velvet: it was as soft as a carpet, but at the
same time perfectly firm to the grip of the boot.
The sun was hidden behind the mountain; and,
thus steeped in shade, we could enjoy, with spirits
unblunted by the heat, the loveliness and grandeur
of the scene.

Right before us was the pyramid of the Aletschhorn, bearing its load of glaciers, and thrusting
above them its pinnacle of rock; while right and
left of us towered and fell to snowy cols such other
peaks as usually hang about a mountain of nearly
14,000 feet elevation. And amid them all, with a
calmness corresponding to the deep seclusion of the
place, wound the beautiful system of glaciers along
which we had been marching for nearly three hours.
I know nothing which can compare in point of
glory with these winter palaces of the mountaineer,
under the opening illumination of the morning.

[1] 'Glaciers of the Alps,' p. 234.

And the best of it is, that no right of property in the scene could enhance its value. To Switzerland belongs the rock—to the early climber, competent to enjoy them, belong the sublimity and beauty of mass, form, colour, and grouping. And still the outward splendour is by no means all. 'In the midst of a puddly moor,' says Emerson, 'I am afraid to say how glad I am:' which is a strong way of affirming the influence of the inner man as regards the enjoyment of external nature. And surely the inner man is a high factor in the effect. The magnificence of the world outside suffices not. Like light falling upon the polished plate of the photographer, the glory of Nature, to be felt, must descend upon a soul prepared to receive its image and superscription.

Mind, like force, is known to us only through matter. Take, then, what hypothesis you will—consider matter as an instrument through which the insulated mind exercises its powers, or consider both as so inextricably mixed that they stand or fall together; from both points of view the care of the body is equally important.[1] The morality of clean blood ought to be one of the first lessons taught us by our pastors and masters. The physical is the substratum of the spiritual, and this fact ought

[1] It will not be supposed that I here mean the stuffing or pampering of the body. The shortening of the supplies, or a good monkish fast at intervals, is often the best discipline for the body.

to give the food we eat and to the air we breathe
a transcendental significance. Boldly and truly
writes Mr. Ruskin, 'Whenever you throw your
window wide open in the morning, you let in
Athena, as wisdom and fresh air at the same instant;
and whenever you draw a pure, long, full breath of
right heaven, you take Athena into your heart, through
your blood; and with the blood into thoughts of
the brain.' No higher value than this could be
assigned to atmospheric oxygen.

Precisely three hours after we had quitted our
hotel the uniform gradient of the Aletschhorn glacier
came to an end. It now suddenly steepened to run
up the mountain. At the base we halted to have
some food, a huge slab of granite serving us for a
table. It is not good to go altogether without food
in these climbing expeditions; nor is it good to eat
copiously. Here a little and there a little, as the
need makes itself apparent, is the prudent course.
For, left to itself, the stomach infallibly sickens, and
the forces of the system ooze away. Should the
sickness have set in so as to produce a recoil from
nutriment, the stomach must be forced to yield.
A small modicum of food usually suffices to set
it right. The strongest guides and the sturdiest
porters have sometimes to use this compulsion.
'Sie müssen sich zwingen.' The guides refer
the capriciousness of the stomach at great eleva-

tions to the air. This may be *a* cause, but I am inclined to think that something is also due to the motion—the long-continued action of the same muscles upon the diaphragm. The condition of things antecedent to the journey must also be taken into account. There is little, if any, sleep; the starting meal is taken at an unusual hour; and if the start be made from a mountain cave or cabin, instead of from the bed of an hotel, the deviation from normal conditions is aggravated. It could not be the mere difference of height between Mont Blanc and Monte Rosa which formerly rendered their effects upon travellers so different. It is that, in the one case, you had the melted snow of the Grands Mulets for your coffee, and a bare plank for your bed; while in the other you had the comparative comforts of the auberge on the Riffel. On the present occasion I had a bottle of milk, which suits me better than anything else. That and a crust are all I need to keep my vigour up and to ward off *le mal des montagnes*.

After half an hour's halt we made ready for the peak, meeting first a quantity of moraine matter mingled with patches of snow, and afterwards the rifted glacier. We threaded our way among the crevasses, and here I paid particular attention to the deportment of my guide. The want of confidence, or rather the absence of that experience of a guide's powers, on which alone perfect reliance can be based, is a serious drawback to the climber. This source of weakness has often come home to me since the death of my brave friend Bennen. His loss to me was like that of an arm to a fighter. But I was glad to notice that my present guide was not likely to err on the score of rashness. He left a wider margin between us and accident than I should have deemed necessary; he sounded with his staff where I should have trod without hesitation; and, knowing my own caution, I had good reason to be satisfied with his. Still, notwithstanding all his vigilance, he once went into a concealed fissure—only waist-deep, however, and he could certainly have rescued himself without the tug of the rope which united us.

After some time we quitted the ice, striking a rocky shoulder of the mountain. The rock had been pulled to pieces by the weather, and its fragments heaped together to an incoherent ridge. Over the lichened stones we worked our way, our course, though rough, being entirely free from danger. On this ridge the sun first found us, striking us at intervals, and at intervals disappearing behind the sloping ridge of the Aletschhorn. We attained the summit of the rocks, and had now the upper reaches of the *névé* before us. To our left the glacier was greatly torn, exposing fine vertical sections, deep blue pits and chasms, which were bottomless to vision; and ledges, from whose copings hung vaster stalactites than those observed below. The beauty of the higher crevasses is mightily enhanced by the long transparent icicles which hang from their eaves, and which, loosened by the sun, fall into them with ringing sound. Above us was the customary Bergschrund; but the spring avalanches had swept over it, and closed it, and since the spring it had not been able to open its jaws. At this schrund we aimed, reached it, and crossed it, and immediately found ourselves at the base of the final cap of the mountain.

Looking at the Aletschhorn from the Sparrenhorn, or from any other point which commands a similar view of the pyramid, we see upon the ridge which falls from the summit to the right, and at a considerable distance from the top, a tooth or pinnacle of rock, which encloses with the ridge a deep indentation. At this gap we now aimed. We varied our ascent from steep snow to rock, and from steep rock to snow, avoiding the difficulties when possible, and facing them when necessary. We met some awkward places, but none whose subjugation was otherwise than pleasant, and at length surmounted the edge of the *arête*. Looking over this, the facette of the pyramid fell almost sheer to the Middle Aletsch glacier. This was a familiar sight to me, for years ago I had strolled over it alone. Below it was the Great

Aletsch, into which the Middle Aletsch flows, and
beyond both was the well-known ridge of the Æggischhorn. We halted, but only for a moment. Turning suddenly to the left, we ascended the rocky ridge
to a sheltered nook which suggested a brief rest and
a slight renewal of that nutriment which, as stated,
isso necessary to the wellbeing of the climber.

From time to time during the ascent I examined
the polarisation of the sky.   I should not have halted
had not the fear of haze or clouds upon the summit
admonished me.   Indeed, as we ascended, one thin,
arrowy cloud shot like a comet's tail through the air
above us, spanning ninety degrees, or more, of the
heavens.   Never, however, have I observed the sky
of a deeper, darker, and purer blue.   It was to examine this colour that I ascended the Aletschhorn,
and I wished to observe it where the hue was deepest
and the polarisation most complete.   You can look
through very different atmospheric thicknesses at
right angles to the solar beams.   When, for example,
the sun is in the eastern or western horizon, you
can look across the sun's rays towards the northern
or southern horizon, or you can look across them to
the zenith.   In the latter direction the blue is deeper
and purer than in either of the former, the proportion of the polarised light of the sky to its total light
being also a maximum.

The sun, however, when I was on the Aletschhorn,

was not in the horizon, but high above it. I
placed my staff upright on a platform of snow.
It cast a shadow. Inclining the staff *from* the
sun, the shadow lengthened for a time, reached its
major limit, and then shortened. The simplest
geometrical consideration will show that the staff
when its shadow was longest was perpendicular to
the solar rays; the atmosphere in this direction was
shallower and the sky bluer than in any other direction perpendicular to the same rays. Along this
line I therefore looked through the Nicol. The light,
I found, could be quenched so as to leave a residue as
dark as the firmament upon a moonless night; but
still there *was* a residue—the polarisation was not
complete. Nor was the colour, however pure its
appearance, by any means a monochromatic blue.
A disc of selenite, gradually thickening from the
centre to the circumference, when placed between the
Nicol and the sky, yielded vivid *iris* colours. The
blue was very marked; but there was vivid purple,
which requires an admixture of red to produce it.
There was also a bright green, and some yellow. In
fact, however purely blue the sky might seem, it sent
to the eye all the colours of the spectrum: it owed
its colour to the *predominance* of blue, that is to
say, to the enfeeblement, and not to the extinction,
of the other colours of the spectrum. The green
was particularly vivid in the portion of the sky

nearest to the mountains, where the light was 'daf-fodil.'

A pocket spectroscope confirmed these results. Permitting the light of an illuminated cloud to enter the slit of the instrument, a vivid spectrum was observed; but on passing beyond the rim of the cloud to the adjacent firmament, a sudden fall in the intensity of all the less refrangible rays of the spectrum was observed. There was an absolute shortening of the spectrum in the direction of the red, through the total extinction of the extreme red. The fall in luminousness was also very striking as far as the green; the blue also suffered, but not so much as the other colours.

The scene as we ascended grew more and more superb, both as regards grouping and expansion. Viewed from the Bel Alp the many-peaked Dom is a most imposing mountain; it has there no competitor. The mass of the Weisshorn is hidden, its summit alone appearing. The Matterhorn, also, besides being more distant, has a portion of its pyramid cut obliquely away by the slope of the same ridge that intercepts the Weisshorn, and which is seen to our right when we face the valley of the Rhone, falling steeply to the promontory called the Nessel. Viewed from this promontory, the Dom finds its match, and more than its match, in its mighty neighbour, whose hugeness is here displayed from top to bottom. On the lower reaches of the Aletschhorn also the Dom maintains its superiority, the Weisshorn being for a time wholly unseen, and the Matterhorn but imperfectly. As we rise, however, the Dom steadily loses its individuality, until from the ridge of the Aletschhorn it is jumbled to a single leviathan heap with the mass of Monte Rosa. The Weisshorn meanwhile as steadily gains in grandeur, rising like a mountain Saul amid the congregated hills, until from the *arête* it distances all competitors. In comparison with this kingly peak, the Matterhorn looks small and mean. It has neither the mass nor the form which would enable it to compete, from a distant point of view, with the Weisshorn.

The ridge of the Aletschhorn is of schistose gneiss, in many places smooth, in all places steep, and sometimes demanding skill and strength on the part of the climber. I thought we could scale it with greater ease if untied, so I flung the rope away from me. My guide was in front, and I carefully watched his action among the rocks. For some time there was nothing to cause anxiety for his safety. There was no likelihood of a slip, and if a slip occurred there was opportunity for recovery. But after a time this ceased to be the case. The rock had been scaled away by weathering parallel to the planes of foliation, the surfaces left behind being excessively smooth, and in many cases flanked by slopes and couloirs of perilous steepness. I saw that a slip might occur here, and that its consequences would be serious. The rope was therefore resumed.

A fair amount of skill and an absence of all precipitancy rendered our progress perfectly secure. In every place of danger one of us planted himself as securely as the rock on which he stood, and remained thus fixed until the danger was passed by the other. Both of us were never exposed to peril at the same moment. The bestowal of a little extra time renders this arrangement possible along the entire ridge of the Aletschhorn; in fact, the dangers of the Alps can be almost reduced to the level of the dangers of the street by the exercise of skill and caution. For rashness, ignorance, or carelessness the mountains leave no margin; and to rashness, ignorance, or carelessness three-fourths of the catastrophes which shock us are to be traced. Even those whose faculties are ever awake in danger are sometimes caught napping when danger seems remote; they receive accordingly the punishment of a tyro for a tyro's neglect.[1]

While ascending the lower glacier we found the air in general crisp and cool; but we were visited at intervals by gusts of Föhn—warm breathings of

---

[1] My own carelessness, as described towards the close of this chapter, may be taken as an illustration.

the unexplained Alpine sirocco, which passed over
our cheeks like puffs from a gently heated stove.
On the *arête* we encountered no Föhn; but the
rocks were so hot as to render contact with them
painful. I left my coat among them, and went
upward in my shirt-sleeves. At our last bivouac
my guide had allowed two hours for the remaining
ascent. We accomplished it in one, and I was surprised by the shout which announced the passage
of the last difficulty, and the proximity of the top
of the mountain. This we reached precisely eight
hours after starting—an ascent of fair rapidity, and
without a single mishap from beginning to end.

Rock, weathered to fragments, constitutes the
crown of the Aletschhorn; but against this and
above it is heaped a buttress of snow, which tapers,
as seen from the Æggischhorn, to a pinnacle of sur-
passing beauty. This snow was firm, and we readily
attained its highest point. Over this I leaned for
ten minutes, looking along the face of the pyramid,
which fell for thousands of feet to the *névés* at its
base. We looked *down* upon the Jungfrau, and
upon every other peak for miles around us, one only
excepted. The exception was the Finsteraarhorn,
the highest of the Oberland mountains, after which
comes the Aletschhorn. I could clearly track the
course pursued by Bennen and myself eleven years
previously—the spurs of rock and slopes of snow,

the steep and weathered crest of the mountain, and the line of our swift glissade as we returned.

Round about the dominant peak of the Oberland was grouped a crowd of other peaks, retreating eastward to Graubünden and the distant Engadin; retreating southward over Italy, and blending ultimately with the atmosphere. At hand were the Jungfrau, Mönch, and Eiger. A little further off the Blumlis Alp, the Weisse Frau, and the Great and Little Nesthorn. In the distance the grim precipices of Mont Blanc, rising darkly from the Allée Blanche, and lifting to the firmament the snow-crown of the mountain. The Combin and its neighbours were distinct; and then came that trinity of grandeur, with which the reader is so well acquainted—the Weisshorn, the Matterhorn, and the Dom—supported by the Alphubel, the Allaleinhorn, the Rympfischhorn, the Strahlhorn, and the mighty Monte Rosa. From no other point in the Alps have I had a greater command of their magnificence—perhaps from none so great; while the blessedness of perfect health, on this perfect day, rounded off within me the external splendour. The sun seemed to take a pleasure in bringing out the glory of the hills. The intermixture of light and shade was astonishing; while to the whole scene a mystic air was imparted by a belt of haze, in which the furthest outlines disappeared, as if infinite distance had rendered them impalpable.

Two concentric shells of atmosphere, perfectly distinct in character, clasped the earth this morning. That which hugged the surface was of a deep neutral tint, too shallow to reach more than midway up the loftier mountains. Upon this, as upon an ocean, rested the luminous higher atmospheric layer, both being separated along the horizon by a perfectly definite line. This higher region was without a cloud; the arrowy streamer that had shot across the firmament during our ascent, first reduced to feathery streaks, had long since melted utterly away. Blue was supreme above, while all round the horizon the intrinsic brilliance of the upper air was enhanced by contrast with the dusky ground on which it rested. But this gloomier portion of the atmosphere was also transparent. It was not a cloud-stratum cutting off the view of things below it, but an attenuated mist, through which were seen, as through a glass darkly, the lower mountains, and out of which the higher peaks and ridges sprung into sudden glory.

Our descent was conducted with the same care and success that attended our ascent. I have already stated it to be a new thing for one man to lead a traveller up the mountain, and my guide in ascending had informed me that his wife had been in a state of great anxiety about him. But until he had cleared all dangers he did not let me know the extent of her devotion, nor the means she had adopted to ensure his safety. When we were once more upon the lower glacier, having left all difficulties behind us, he remarked with a chuckle that she had been in a terrible state of fear, and had informed him of her intention to have a mass for his safety celebrated by the village priest. But if he profited by this mediation, I must have done so equally; for in all dangerous places we were tied together by a rope which was far too strong to break had I slipped. My safety was, in fact, bound up in his, and I therefore thought it right to pay my share of the expense. 'How much did the mass cost?' I asked. 'Oh, not much, sir,' he replied; 'only ninety centimes.' Not deeming the expense worth dividing, I let him pay for such advantage as I had derived from the priest's intercession.

In 1868 I had been so much broken down on going to the Alps that even amongst them I found it difficult to recover energy. In 1869, however, after a severe discipline in bathing and climbing,[1] my weariness disappeared, and before I attacked the Aletschhorn I felt that my restoration was ensured. In my subsequent rambles it was a great

[1] In 1869 I tried to get to the top of the Wetterhorn in a single day from Grindelwald, but the wildness of the storm and the bitterness of the cold drove Peter Baumann and me back, when we were within a quarter of an hour of the top. I was afterwards in the habit of taking to the Riffel See when heavy snow was falling. It was at the Bel Alp, however, that I found myself renewed.

delight and refreshment to me, whenever I felt heated, to choose a bubbling pool in some mountain stream, roll myself in it, and afterwards dance myself dry in the sunshine. Each morning I had a tub in a rivulet, a header in a lake, or a douche under a cascade. The best of these was half a mile or more from the hotel, but there was an inferior waterfall close at hand to which I resorted when time was short. On a bright morning towards the end of August 1869 I was returning from this cascade to my clothes, which were about twenty yards off. They might have been reached by walking on the grass, but I chose to walk on some slippery blocks of gneiss, and using no caution I staggered and fell. My shin was urged with great force against the sharp crystals, which inflicted three ugly wounds; but I sponged the blood away, wrapped a cold bandage round the injured place, and limped to the hotel. I was quite disabled, but felt sure of speedy recovery, my health was so strong.

For four or five days I remained quietly in bed. The wound had become entirely painless; there was hardly any inflammation and no pus. I felt so well that I thought a little exercise would do me less harm than good. I abandoned my cold bandage and went out. That night inflammation set in, pus appeared, and in trying to dislodge it I poisoned the wound. It became worse and worse;

318 HOURS OF EXERCISE IN THE ALPS.   [1869

erysipelas set in, and at last it became evident that I might lose my foot or something more important. After remaining nearly a fortnight at the Bel Alp without medical advice, I resolved to go to Geneva. I wrote accordingly to my friend Professor De la Rive, with the view of securing the services of an able surgeon. I was carried down to Brieg on a kind of bier, and midway on the mountain-slope had the good fortune to meet Mr. Ellis of Sloane Street. He examined my wound, and I have good reason to feel grateful to him for his extreme kindness and his excellent advice. My friend Soret met me at the railway station, and Dr. Gauthier was at my side a few seconds after I entered my hotel.

But, despite all the care, kindness, and real skill bestowed upon me, I was a month in bed at Geneva. A sinus about five inches long had worked its channel from the wound down to the instep, which was undermined by an abscess. This Dr. Gauthier discovered, and by assiduous attention cured. In her beautiful residence at Lammermor, on the margin of Lake Leman, Lady Emily Peel had a bed erected for me as soon as I was able to go there, and it was under her roof that the last traces of the sinus disappeared. I was so emaciated, however, that it required several months to restore the flesh and the strength that this paltry accident cost me.

In 1870 I was again at the Bel Alp for several
weeks, during which my interest was continually
kept awake by telegrams from the seat of war; for
the enterprising proprietors both at the Bel Alp and
the Æggischhorn had run telegraphic wires from
the valley of the Rhone up to their respective hotels.
The most noteworthy occurrence among the mountains in 1870 was a terrific thunderstorm, which set
two forests on fire by the same discharge.   One fire
near the Rieder Alp was speedily quenched; the
other, under the Nessel, burned for several successive
days and nights, and threatened to become a public
calamity. A constant fiery glow was kept up by the
combustion of the underwood, which formed the vehicle of transmission among the larger trees.   Three
or four of these would often burst simultaneously
into pyramids of flame, which would last but a few
minutes, leaving the trees with all their branches as
red-hot embers behind.   Heavy and persistent rain
at length extinguished the conflagration.

XXVI.

A DAY AMONG THE SÉRACS OF THE GLA-
CIER DU GÉANT FOURTEEN YEARS
AGO.

HAVING fixed my head-quarters at the Montanvert,
I was engaged for nearly six weeks during the
summer of 1857 in making observations on the Mer
de Glace and its tributaries. Throughout this time
I had the advantage of the able and unremitting
assistance of my friend Mr. Hirst, who kindly undertook, in most cases, the measurement of the motion
of the glacier. My permanent guide, Édouard
Simond, an intelligent and trustworthy man, was
assistant on these occasions, and having arranged
with Mr. Hirst the measurements required to be
made, it was my custom to leave the execution of
them to him, and to spend much of my time alone
upon the glaciers. Days have thus been occupied
amid the confusion of the Glacier du Géant, at the
base of the great ice-fall of La Noire, in trying
to connect the veined structure of the glacier with

the stratification of its *névé*; and often, after wandering almost unconsciously from peak to peak and from hollow to hollow, I have found myself, as the day was waning, in places from which it required a sound axe and a vigorous stroke to set me free.

This practice gradually developed my powers of dealing with the difficulties of the glacier.   On some occasions, however, I found the assistance of a companion necessary, and it was then my habit to take with me a hardy boy named Balmat, who was attached to the hotel at the Montanvert.   He could climb like a cat, and one of our first expeditions together was an ascent to a point above Trélaporte, from which a magnificent view of the entire glacier is obtained.   This point lies under the Aiguille de Charmoz, and to the left of a remarkable cleft, which is sure to attract the traveller's attention on looking upwards from the Montanvert.   We reached the place through a precipitous couloir on the Montanvert side of the mountain; and while two chamois watched us from the crags above, we made our observations, and ended our survey by pledging the health of Forbes and other explorers of the Alps.

We descended from the eminence by a different route ; during both ascent and descent I had occasion to admire the courage and caution of my young companion, and the extraordinary cohesive

Y

force by which he clung to the rock. He, moreover, evidently felt himself responsible for my safety, and once when I asserted my independence so far as to attempt descending a kind of 'chimney,' which, though rather dangerous-looking, I considered to be practicable, he sprang to my side, and, with out-stretched arm and ringing voice, exclaimed, 'Monsieur, je vous défends de passer par là !'

Anxious to avoid the inconvenience of the rules of the Chamouni guides, my aim, from the first, was to render myself as far as possible independent of their assistance. Wishing to explore the slopes of the Col du Géant, not for the purpose of crossing into Piedmont, but to examine the fine ice-sections which it exhibits, and to trace amid its chasms the gradual conversion of the snow into ice, I at first thought of attempting the ascent of the col alone ; but 'le petit Balmat,' as my host at the Montanvert always named him, acquitted himself so well on the occasion referred to that I thought he would make a suitable companion. On naming the project to him he eagerly embraced my proposal ; in fact, he said he was willing to try Mont Blanc with me if I desired it.

On the morning of Friday, July 24, we accordingly set off for the Tacul, I making, as we ascended, such few observations as lay in our way. The sun shone gloriously upon the mountains, and gleamed by reflection from the surface of the glacier. Looked at through a pair of very dark spectacles, the scene was exceedingly striking and instructive. Terraces of snow clung to the mountains, exposing, here and there, high vertical sections, which cast dense shadows upon the adjacent plateaux. The glacier was thrown into heaps and 'hummocks,' their tops glistening with white, silvery light, and their sides intensely shaded. When the lateral light was quite shut out, and all that reached the eyes had to pass through the spectacles, the contrast between light and shade was much stronger than when the glacier was viewed by the broad light of day. In fact, the shadows were no longer grey merely, but black; to a similar augmentation of contrast towards the close of the day is to be referred the fact that the 'Dirt Bands' of the Mer de Glace are best seen by twilight.

A gentleman had started in the morning to cross the col, accompanied by two strong guides. We met a man returning from the Jardin, who told us that he had seen the party that preceded us; that they had been detained a long time amid the séracs, and that our ascending without ladders was quite out of the question. As we approached the Tacul, my lynx-eyed little companion ranged with the telescope over the snowy slopes of the col, and at length exclaimed, 'Je les vois, tous les trois!'—

the 'Monsieur' in the middle, and a guide before and behind. They seemed like three black specks upon the shoulders of the Giant; below them was the vast ice-cascade, resembling the foam of ten Niagaras placed end to end and stiffened into rest, while the travellers seemed to walk upon a floor as smooth as polished Carrara marble. Here and there, however, its uniformity was broken by vertical faults, exposing precipices of the stratified *névé*.

On an old moraine near the Tacul, piled up centuries ago by the Glacier de Léchaud, immense masses of granite are thrown confusedly together; and one enormous slab is so cast over a number of others as to form a kind of sheltered grotto, which we proposed to make our resting-place for the night. Having deposited our loads here, I proceeded to the icefall of the Talèfre, while my companion set out towards the Couvercle in search of firewood. I walked round the base of the frozen cascade, and climbed up among its riven pinnacles, examining the structure as I ascended. The hollow rumble of the rocks as they fell into the crevasses was incessant. From holes in the ice-cliffs clear cataracts gushed, coming I knew not whence, and going I knew not whither. Sometimes the deep gurgle of sub-glacial water was heard, far down in the ice. The resonance of the water as it fell into shafts struck me suddenly at intervals on turning corners, and seemed, in each

case, as if a new torrent had bounded into life.
Streams flowed through deep channels which they
themselves had worn, revealing beautifully the 'rib-
boned structure.' At the further end of the Glacier
de Léchaud the Capucin Rock stood, like a preacher:
and below him a fantastic group of granite pinnacles
suggested the idea of a congregation. The outlines
of some of the ice-cliffs were also very singular ;
and it needed but a slight effort of the imagination
to people the place with natural sculpture.

At six o'clock the shrill whistle of my companion
announced that our time of meeting was come.
He had found some wood—dry twigs of rhododen-
drons, and a couple of heavy stumps of juniper. I
shouldered the largest of the latter, while he
strapped his twigs on his back, and led the way to
the Tacul. The sun shot his oblique rays against us
over the heights of Charmoz, and cast our shadows
far up the glacier. We filled our saucepan, which
Balmat named ' a machine,' with clear water, and
bore it to our cavern, where the fire was soon
crackling under the machine. I was assailed by the
smoke, which set my eyes dripping tears ; but this
cleared away when the fire brightened, and we
boiled our chocolate and made a comfortable evening
meal.

I afterwards clambered up the moraine to watch
the tints of the setting sun ; clouds floated round the

Aiguille de Charmoz, and were changed from grey to red, and from red to grey, as their positions varied. The shadows of the isolated peaks and pinnacles were drawn, at times, in black bands across the clouds; and the Aiguille du Moine smiled and frowned alternately. One high snow-peak alone enjoyed the unaltered radiance of the sinking day; the sunshine never forsook it, but glowed there, like the steady light of love, while a kind of coquetry was carried on between the atmosphere and the surrounding mountains. The notched summits of the Grande and Petite Jorasse leaned peacefully against the blue firmament. The highest mountain-crags were cleft, in some cases, into fantastic forms; single pillars stood out from all else, like lonely watchers, over the mountain scene; while little red clouds playfully embraced them at intervals, and converted them into pillars of fire.

The sun at length departed, and all became cold and grey upon the mountains; but a brief secondary glow came afterwards, and warmed up the brown cliffs once more. I descended the moraine, the smell of the smoke guiding me towards the rock under which I was to pass the night. A fire was burning at the mouth of the grotto, reddening with its glare the darkness of the interior. Beside the fire sat my little companion, with a tall, conical, red night-cap drawn completely over his ears; our saucepan

was bubbling on the fire; he watched it medita-
tively, adding at times a twig, which sprung immediately into flame, and strengthened the glow
upon his countenance. He looked, in fact, more like
a demon of the ice-world than a being of ordinary
flesh and blood. I had been recommended to take
a bit of a tallow candle with me to rub my face
with, as a protection against the sun; by the light
of this we spread our rugs, lay down upon them, and
wrapped them round us.

The countless noises heard upon the glacier during
the day were now stilled, and dead silence ruled the
ice-world; the roar of an occasional avalanche, however, shooting down the flanks of Mont Mallet, broke
upon us with startling energy. I did not sleep till
towards four o'clock in the morning, when I dozed
and dreamed, and mingled my actual condition with
my dream. When I awoke, I found my head weary
enough upon the clay of the old moraine, my ribs
pressed closely against a block of granite, and my
feet amid sundry fragments of the same material.
It was nearly five o'clock on Saturday the 25th
when I arose; my companion quickly followed my
example. He also had slept but little, and once or
twice during the night I fancied I could feel him
shiver. We were, however, well protected from the
cold. The high moraine of the Glacier du Léchaud
was on one side, that of the Glacier du Géant on

the other, while the cliffs of Mont Tacul formed the
third side of a triangle, which sheltered us from the
sharper action of the wind. At times the calm was
perfect, and I felt almost too warm; then again a
searching wind would enter the grotto, and cause
the skin to shrink on all exposed parts of the body.
It had frozen hard, and to obtain water for washing
I had to break through a sheet of ice which coated
one of the pools upon the glacier.

In a few minutes our juniper fire was crackling
cheerily; we made our chocolate and breakfasted.
My companion emptied the contents of a small
brandy bottle into my flask, which, however, was
too small to hold it all, and on the principle, I
suppose, of avoiding waste, he drank what remained.
It was not much, but sufficient to muddle his brain,
and to make him sluggish and drowsy for a time.
We put the necessary food in our knapsacks and
faced our task, first ascending the Glacier du
Tacul along its eastern side, until we came to the
base of the séracs.

The vast mass of snow collected on the plateau
of the Col du Géant, and compressed to ice by its
own weight, reaches the throat of the valley, which
stretches from the rocks called Le Rognon to the
promontory of the Aiguille Noire. Through this
defile it is forced, falling steeply, and forming one
of the grandest ice-cascades in the Alps. At the

summit it is broken into transverse chasms of enormous width and depth ; the ridges between these
break across again, and form those castellated
masses to which the name of *séracs* has been
applied. In descending the cascade the ice is
crushed and riven; ruined towers, which have
tumbled from the summit, cumber the slope, and
smooth vertical precipices of ice rise in succession
out of the ruins. At the base of the fall the fragments are again squeezed together, but the confusion is still great, the glacier being tossed into
billowy shapes, scooped into caverns, and cut into
gorges by torrents which expand here and there
into deep green lakes.

Across this portion of the glacier we proceeded
westward, purposing to attempt the ascent at the
Rognon side.[1] Perils and difficulties soon began to
thicken round us. The confusion of ice-pinnacles,
crags, and chasms was very bewildering. Plates of

[1] Standing here alone, on another occasion, I heard the roar of
what appeared to be a descending avalanche, but the duration of the
sound suprised me. I looked through my opera-glass in the direction
from which the sound proceeded, and saw issuing from the end of
one of the secondary glaciers on the side of Mont Tacul a torrent of
what appeared to me to be stones and mud. I could see the rocks
and *débris* jumping down the declivities, and forming singular
cascades. The noise continued for a quarter of an hour, when the
descending torrent diminished until the ordinary stream, due to
the melting of the glacier, alone remained. A sub-glacial lake had
evidently burst its bounds, and carried the *débris* along with it in its
rush downwards.

ice jutted from the glacier like enormous fins, up
the sides of which we had to rise by steps, and along
the edges of which we had to walk. Often, while
perched upon these eminences, we were flanked
right and left by crevasses, the depth of which
might be inferred from their impenetrable gloom.
At some places forces of extreme complexity had
acted on the mass; the ridges were broken into
columns, and some of these were twisted half round;
while the chasms were cut up into shafts which
resembled gigantic honeycombs. Our work was
very difficult, sometimes disheartening : nevertheless, our inspiration was, that what man has done
man may do, and we accordingly persevered. My
fellow-traveller was silent for a time : the brandy
had its effect upon him, and he confessed it; but
I thought that a contact with the cold ice would
soon cause this to disappear, after which I resolved
not to influence his judgment in the least.

Looking now to the right, I suddenly became
aware that, high above us, a multitude of unstable
crags and leaning columns of ice covered the precipitous incline. We had reached a position where
protecting cliffs rose to our right, while in front of
us was a space more open than any we had yet
passed. The reason was that the ice avalanches had
chosen it for their principal path. We had stepped
upon this space when a peal above us brought us to

a stand. Crash! crash! crash! nearer and nearer,
the sound becoming more continuous and confused,
as the descending masses broke into smaller blocks.
Onward they came! boulders half a ton and more
in weight, leaping down with a kind of maniacal
fury, as if their sole mission was to crush the séracs
to powder. Some of them on striking the ice
rebounded like elastic balls, described parabolas
through the air, again smote the ice, and scattered
its dust like clouds in the atmosphere. Deflected
by their collision with the glacier, some blocks were
carried past us within a few yards of the spot where
we stood. I had never before witnessed an exhibition of force at all comparable to this, and its
proximity rendered that fearful which at a little
distance would have been sublime.

My companion held his breath, and then exclaimed, 'C'est terrible! il faut retourner.' In
fact, while the avalanche continued we could not
at all calculate upon our safety. When we heard
the first peal we had instinctively retreated to the
shelter of the ice bastions; but what if one of these
missiles struck the tower beside us! would it be
able to withstand the shock? We knew not. In
reply to the proposal of my companion, I simply
said, 'By all means, if you desire it; but let us
wait a little.' I felt that fear was just as bad a
counsellor as rashness, and thought it but fair to

wait until my companion's terror had subsided. We waited accordingly, and he seemed to gather courage and assurance. I scanned the heights and saw that a little more effort in an upward direction would place us in a much less perilous position, as far as the avalanches were concerned. I pointed this out to my companion, and we went forward. Once indeed, for a minute or two, I felt anxious. We had to cross in the shadow of a tower of ice, of a loose and threatening character, which quite over-hung our track. The freshly broken masses at its base, and at some distance below it, showed that it must have partially given way some hours before. ' Don't speak or make any noise,' said my companion ; and, although rather sceptical as to the influence of speech in such a case, I held my tongue and escaped from the dangerous vicinity as fast as my legs and alpenstock could carry me.

Unbroken spaces, covered with snow, now began to spread between the crevasses ; these latter, how-ever, became larger, and were generally placed end to end *en échelon*. When, therefore, we arrived at the edge of a chasm, by walking along it we usually soon reached a point where a second one joined on it. The extremities of the chasms ran parallel to each other for some distance, one being separated from the other, throughout this distance, by a wall of incipient ice, coped at the top by snow. At other places, however, the lower portion of the partition between the fissures had melted away, leaving the chasm spanned by a bridge of snow, the capacity of which to bear us was often a matter of delicate experiment.   Over these bridges we stepped as lightly as possible : 'Allez doucement ici,' was the perpetual admonition of my companion, 'et il faut toujours sonder.'

In many cases, indeed, we could not at all guess at the state of matters underneath the covering of snow. We had picked up a few hints upon this subject, but neither of us was at this time sufficiently experienced to make practical use of them. The 'sounding' too was rather weary work, as, to make it of any value, the bâton must be driven into the snow with considerable force. Further up in the *névé* the fissures became less frequent, but some of them were of great depth and width. On those silent heights there is something peculiarly solemn in the aspect of the crevasses, yawning gloomily day and night, as if with a never-satisfied hunger. We stumbled on the skeleton of a chamois, which had probably met its death by falling into a chasm, and been disgorged lower down. But a thousand chamois between these cavernous jaws would not make a mouthful.   I scarcely knew which to choose—these pitfalls of the *névé*, or the avalanches. The latter are terrible, but they are

grand, outspoken things; the ice crags proclaim
from their heights, 'Do not trust us; we are momentary and merciless.' They wear the aspect of
hostility undisguised; but these chasms of the *névé*
are typified by the treachery of the moral world,
hiding themselves under shining coverlets of snow,
and compassing their ends by dissimulation.

After some time we alighted on the trace of those
who had crossed the day before. The danger was
over when we made the discovery, but it saved us
some exploring amid the crevasses which still remained. We at length got quite clear of the fissures,
and mounted zigzag to the summit of the col.
Clouds drove up against us from the valley of
Courmayeur, but they made no way over the col.
At the summit they encountered a stratum of drier
air, mixing with which they were reduced, as fast as
they came, to a state of invisible vapour. Upon the
very top of the col I spread my plaid, and with the
appetites of hungry eagles we attacked our chicken
and mutton. I examined the snow and made some
experiments on sound; but little Balmat's feet were
so cold that he feared being frostbitten, and at his
entreaty we started on our descent again as soon as
possible.

To the top of the séracs we retraced the course by
which we had ascended, but here we lost the track,
for there was no snow to retain it. A new lesson

was before us. We kept nearer to the centre of the
glacier than when we ascended, thereby avoiding the
avalanches, but getting into ice more riven and dislocated. We were often utterly at a loss how to proceed. My companion made several attempts to regain
the morning's track, preferring to risk the avalanches
rather than be blocked and ditched up in an ice-
prison from which we saw no means of escape.
Wherever we turned peril stared us in the face ; but
the recurrence of danger had rendered us callous to
it, and this indifference gave a mechanical surety to
the step in places where such surety was the only
means of avoiding destruction. Once or twice, while
standing on the summit of a peak of ice, and looking
at the pits and chasms beneath me, at the distance
through which we had hewn our way, and at the
work still to be accomplished, I experienced an incipient flush of terror. But this was immediately
drowned in action. Indeed the case was so bad, the
necessity for exertion so paramount, that the will
acquired an energy which crushed out terror. We
proceeded, however, with the most steady watch-
fulness. When we arrived at a difficulty which
seemed insuperable, we calmly inspected it, looking
at it on all sides ; and though we had often to
retrace our steps amid cliffs and chasms, still formidable obstacles repeatedly disappeared before
our cool and searching examination. We made no

haste, we took no rest, but ever tended downwards.
With all our instincts of self-preservation awake, we
crossed places which, without the spur of necessity
to drive us, we should have deemed impassable.

Once, having walked for some distance along the
edge of a high wedge of ice, we had to descend its
left face in order to cross a crevasse. The ice was
of that loose granular character which causes it to
resemble an aggregate of little polyhedrons jointed
together more than a coherent solid. I was not
aware that the substance was so utterly disintegrated
as it proved to be. To aid me in planting my foot
securely on the edge of the crevasse, I laid hold of
a projecting corner of the ice. It crumbled to
pieces in my hand; I tottered for a moment in the
effort to regain my balance, my footing gave way,
and I went into the chasm. I heard my companion
scream, 'O! mon Dieu, il est perdu!' but a ledge
about two feet wide jutted from the side of the
crevasse; and this received me, my fall not amount-
ing to more than three or four feet. A block
of ice which partially jammed up the chasm concealed me from Balmat. I called to him, and he
responded by another exclamation, 'O! mon Dieu,
comme j'ai peur!' He helped me up, and, looking
anxiously in my face, demanded 'N'avez-vous pas
peur?' Afterwards the difficulties lessened by
degrees, and we began to gladden ourselves by

mutual expressions of 'content' with what we had
accomplished. We at length reached the base of
the séracs; ordinary crevasess were trivial in comparison with those from which we had escaped, so
we hastened along the glacier, without halting, to
the Tacul.

Here a paltry accident caused me more damage
than all the dangers of the day. I was passing
a rock, the snow beside it seemed firm, and I
placed my bâton upon it, leaning trustfully upon
the staff. Through the warmth of the rock, or
some other cause, the snow had been rendered
hollow underneath; it yielded, I fell forward, and
although a cat-like capacity of helping myself in
such cases saved me from serious hurt, it did not
prevent my knee from being urged with all my
weight against an edge of granite. I rested for half
an hour in our grotto at the Tacul, and afterwards
struggled lamely along the Mer de Glace home to
the Montanvert. Bloodshot eyes, burnt cheeks, and
blistered lips were the result of the journey, but
these soon disappeared, and fresh strength was
gained for further action.

The above account was written on the day following the ascent, and while all its incidents were fresh
in my memory. Last September, guided by the
tracks of previous travellers, I ascended nearly to the
summit of the ice-fall, along its eastern side, and to

z

those acquainted only with such dangers as I then experienced the account which I have just given must appear exaggerated.   I can only say that the track which I pursued in 1858 bore no resemblance in point of difficulty to that which I followed in 1857.   The reason probably is, that in my first expedition neither myself nor my companion knew anything of the route, and we were totally destitute of the adjuncts which guides commonly use in crossing the ' Grand Col.'

NOTES AND COMMENTS

ICE AND GLACIERS

AND OTHER SCRAPS.

VOYAGE TO ALGERIA TO OBSERVE
THE ECLIPSE.

I.

OBSERVATIONS ON THE MER DE GLACE.

THE law established by Forbes and Agassiz, that the central portions of a glacier moved faster than the sides, was amply illustrated and confirmed by the deportment of lines of stakes placed across the Mer de Glace and its tributaries in 1857. The portions of the trunk glacier derived from these tributaries were easily traceable throughout the glacier by means of the *moraines*. Thus, for example, the portion of the trunk stream derived from the Glacier du Géant might be distinguished in a moment from the other portions by the absence of *débris* upon its surface. Attention was drawn by Prof. Forbes to the fact that the eastern side of the Mer de Glace in particular is 'excessively crevassed;' and he accounted for this crevassing by supposing that the Glacier du Géant moves most swiftly, and in its effort to drag its more sluggish companions along with it tears them asunder, thus producing the fissures and dislocation for which the eastern side of the glacier is remarkable. Too much weight must not be attached to this explanation.

It was one of those suggestions which are perpetu-
ally thrown out by men of science during the
course of an investigation, and the fulfilment or
non-fulfilment of which cannot materially affect
the merits of the investigator. Indeed, the merits
of Forbes must be judged on far broader grounds.
The qualities of mind and the physical culture
invested in his 'Travels in the Alps' are such as
to make it, in the estimation of the physical investigator at least, outweigh all other books upon
the subject.

While thus acknowledging its merits, however,
let a free and frank comparison of its statements
with facts be instituted. To test whether the
Glacier du Géant moved more quickly than its
fellows, *five different lines* were set out across the
Mer de Glace, in the vicinity of the Montanvert.
In each case it was found that the point of swiftest
motion did not lie upon the Glacier du Géant at
all, but was displaced so as to bring it comparatively close to the eastern side of the glacier. But
though the special opinion of Forbes just referred
to here falls to the ground, the deviation of the
point of swiftest motion from the centre of the
glacier will probably, when its cause is pointed out,
be regarded as of special importance to his theory.

At the place where these five lines were run
across it the glacier turns its convex curvature to

the eastern side of the valley, being concave towards
the Montanvert. Let us then take a bolder analogy
than even that suggested in the explanation of
Forbes, where he compares the Glacier du Géant to
a strong and swiftly-flowing river. Let us enquire
how a river would behave in sweeping round a curve
similar to that here existing. The point of swiftest
motion would undoubtedly lie on that side of the
centre of the stream towards which it turns its
convex curvature. Can this be the case with the
trunk of the Mer de Glace? If so, then we ought
to have a shifting of the point of maximum motion
towards the eastern side of the valley, when the
curvature of the glacier so changes as to turn its
convexity to the western side.

Now, such a change of flexure actually occurs
opposite the passages called *Les Ponts*, and at this
place the view just enunciated was tested. It was
immediately ascertained that the point of swiftest
motion here lay at a different side of the axis from
that observed lower down. But to confer strict
numerical accuracy upon the result, stakes were
fixed at certain distances from the western side of
the glacier, and others *at equal distances* from the
eastern side. The velocities of these stakes were
compared with each other, two by two, a stake on
the western side being always compared with a
second one which stood at the same distance from

the eastern side. The results of this measurement are given in the following table, the numbers denoting inches:

	1st pair	2nd pair	3rd pair	4th pair	5th pair
West .	15	$17\frac{1}{4}$	$22\frac{1}{4}$	$23\frac{3}{4}$	$23\frac{3}{4}$
East .	$12\frac{1}{2}$	$15\frac{1}{4}$	$15\frac{1}{2}$	$18\frac{1}{4}$	$19\frac{1}{2}$

It is here seen that in each case the *western* stake moved more swiftly than its eastern fellow stake; thus proving, beyond a doubt, that opposite the Ponts the western side of the Mer de Glace moves swiftest—a result precisely the reverse of that observed where the curvature of the valley was different.

But an additional test of the explanation is possible. Between the Ponts and the promontory of Trélaporte the glacier passes another point of contrary flexure, its convex curvature opposite to Trélaporte being turned towards the base of the Aiguille du Moine, on the eastern side. A series of stakes was placed across the glacier here; and the velocities of those placed at certain distances from the western side were compared, as before, with those of stakes placed at the same distances from the eastern side. The following table shows the result of these measurements; the numbers, as before, denote inches:

	1st pair	2nd pair	3rd pair
West . .	$12\frac{3}{4}$	15	$17\frac{1}{4}$
East . .	$14\frac{3}{4}$	$17\frac{1}{2}$	19

Here we find that in each case the *eastern* stake moved faster than its fellow. The point of maximum motion has therefore once more crossed the axis of the glacier.

Determining the point of maximum motion for a great number of transverse sections of the Mer de Glaçe, and uniting these points, we have what is called the *locus* of the point. The dotted line in the annexed figure represents the centre of the Mer de Glace; the hard line which crosses the axis of the glacier at the points A A is then the locus of the point of swiftest motion. It is a curve

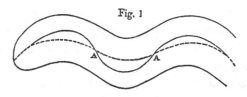

Fig. 1

more deeply sinuous than the valley itself, and it crosses the central line of the valley at each point of contrary flexure. The position of towns upon the banks of rivers is usually on the convex side of the stream, where the rush of the water renders silting-up impossible; and the same law which regulated the flow of the Thames, and determined the position of the towns upon its banks, is at this moment operating with silent energy among the Alpine glaciers.

Another peculiarity of glacier motion is now to be noticed.

Before any observations had been made upon the subject, it was surmised by Prof. Forbes that the portions of a glacier near its bed were retarded by friction against the latter. This view was afterwards confirmed by his own observations, and by those of M. Martins. Nevertheless the state of our knowledge upon the subject rendered further confirmation of the fact highly desirable. A rare opportunity for testing the question was furnished in 1857 by an almost vertical precipice of ice, constituting the side of the Glacier du Géant, exposed near the Tacul. The precipice was about 140 feet in height. At the top and near the bottom stakes were fixed, and by hewing steps in the ice I succeeded in fixing a stake in the face of the precipice at a point about forty feet above the base.[1] After the lapse of a sufficient number of days, the progress of the three stakes was measured ; reduced to the diurnal rate, the motion was as follows :

Top stake	.	.	6·00 inches.
Middle stake	.	.	4·59 ,,
Bottom stake	.	.	2·56 ,,

We thus see that the top stake moved with more

---

[1] It was here that my prudent guide, Édouard Simon, demanded. 'Est-ce que vous avez une femme ?' and, when I replied in the negative, added, 'Vous serez tué tout de même.'

than twice the velocity of the bottom one, while the
velocity of the middle stake lies between the two.
But it also appears that the augmentation of velocity
upwards is not proportional to the distance from the
bottom, but increases in a quicker ratio. At a
height of 100 feet from the bottom, the velocity
would undoubtedly be practically the same as at
the surface. Measurements made upon an adjacent
ice-cliff proved this. We thus see the perfect
validity of the reason assigned by Forbes for the
continued verticality of the walls of transverse
crevasses. Indeed a comparison of the result with
his anticipations and reasonings will prove alike
their sagacity and their truth.

The most commanding view of the Mer de Glace
and its tributaries is obtained from a point above
the remarkable cleft in the mountain-range underneath the Aiguille de Charmoz, which is sure to
attract the attention of an observer standing at the
Montanvert. This point, marked G on the map of
Forbes, I succeeded in attaining. A Tübingen
Professor once visited the glaciers of Switzerland,
and seeing these apparently rigid masses enclosed in
sinuous valleys, went home and wrote a book, flatly
denying the possibility of their motion. An inspection from the point now referred to would have
doubtless confirmed him in his opinion ; and indeed
nothing can be more calculated to impress the

mind with the magnitude of the forces brought into play than the squeezing of the three tributaries of the Mer de Glace through the neck of the valley at Trélaporte.

But let me state numerical results. Previous to its junction with its fellows, the Glacier du Géant measures 1,134 yards across. Before it is influenced by the thrust of the Talèfre, the Glacier de Léchaud has a width of 825 yards; while the width of the Talèfre branch across the base of the cascade, before it joins the Léchaud, is approximately 638 yards. The sum of these widths is 2,597 yards. At Trélaporte those three branches are forced through a gorge 893 yards wide, with a central velocity of 20 inches a day! The result is still more astonishing if we confine our attention to one of the tributaries—that of the Léchaud. This broad ice-river, which before its junction with the Talèfre has a width of 825 yards, at Trélaporte is squeezed to a driblet of less than 88 yards in width, that is to say, to about one-tenth of its previous horizontal transverse dimension.

Whence is the force derived which drives the glacier through the gorge? No doubt pressure from behind. Other facts also suggest that the Glacier du Géant is throughout its length in a state of forcible longitudinal compression. Taking a series of points along the axis of this glacier—if

these points, during the descent of the glacier,
preserved their distances asunder perfectly constant,
there could be no longitudinal compression. The
mechanical meaning of this term, as applied to a
substance capable of yielding like ice, must be that
the hinder points are incessantly advancing upon
the forward ones. I was particularly anxious to test
this view, which first occurred to me on à priori
grounds. Three points, A, B, C, were therefore
fixed upon the axis of the Glacier du Géant, A being
the highest up the glacier. The distance between
A and B was 545 yards, and that between B and C
was 487 yards. The daily velocities of these three
points, determined by the theodolite, were as follows :

A . . . 20·55 inches.
B . . . 15·43   „
C . . . 12·75   „

The result completely corroborates the foregoing
anticipation. The hinder points are incessantly
advancing upon those in front, and that to an
extent sufficient to shorten a segment of this glacier,
measuring 1,000 yards in length, at the rate of
8 inches a day. Were this rate uniform at all
seasons, the shortening would amount to 240 feet in
a year. When we consider the compactness of this
glacier, and the uniformity in the width of the
valley which it fills, this result cannot fail to excite

surprise ; and the exhibition of force thus rendered manifest must be mainly instrumental in driving the glacier through the jaws of the granite vice at Trélaporte.

When the Glacier du Géant is observed from a sufficient distance, a remarkable system of seams of white ice appears to sweep across it, in the direction of the 'dirt-bands.' These seams are more resistant than the ordinary ice of the glacier, and sometimes protrude above the surface to a height of three or four feet. Their origin was for some time a difficulty, and it was at the base of the ice-cascade which descends from the basin of the Talèfre that the key to their solution first presented itself. It was well known that the ice of a glacier is not of homogeneous structure, but that the general more or less milky mass is traversed by blue veins of a more compact and transparent texture. In the upper portions of the Mer de Glace these veins sweep across the glacier in gentle curves, leaning forward—to which leaning forward Prof. Forbes gave the name of the 'frontal dip.' A case of 'backward dip' has never been described. But at the base of the ice-cascade referred to I had often noticed the veins exposed upon the walls of a longitudinal crevasse leaning backwards and forwards on both sides of a vertical line, like the joints of stones used to turn an arch.

This fact was found to connect itself in the following way with the general state of the glacier. At the base of the ice-fall a succession of protuberances, with steep frontal slopes, followed each other, and were intersected by crevasses. Let the hand be placed flat upon the table, with the palm downwards ; let the fingers be bent so as to render the space between the joints nearest the nails and the ends of the fingers nearly vertical. Let the second hand be now placed upon the back of the first, with its fingers bent as in the former case, and their ends resting upon the roots of the first fingers. The crumpling of the hands fairly represents the crumpling of the ice, and the spaces between the fingers represent the crevasses by which the protuberances are intersected. On the walls of these crevasses the change of dip of the veined structure above referred to was always observed, and *at the base of each protuberance a vein of white ice was found firmly wedged into the mass of the glacier.*

The next figure represents a series of these crumples with the veins of white ice *i i i* at their bases.

It was soon observed that the water which trickled down the protuberances, and gushed here and there from glacier orifices, collected at the bases of the crumples, and formed streams which cut for themselves deep channels in the ice. These streams seemed to be the exact matrices or moulds of the

veins of white ice, and the latter were finally traced
to the gorging up of the channels of glacial rivulets
by winter snow. The same explanation applies to
the system of bands upon the Glacier du Géant. I
was enabled to trace the little arms of white ice
which once were the tributaries of the streams, to
see a trunk vein of the ice dividing into branches,
and uniting again so as to enclose glacial islands. I
finally traced them to the region of their formation,

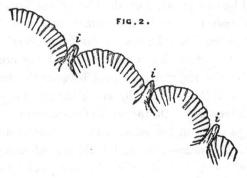

FIG. 2.

and by sketches of existing streams taken near the
base of the *séracs*, and of bands of white ice taken
lower down, a resemblance so striking was exhibited
as to leave no doubt of their relationship. On the
walls of some deep crevasses, moreover, which
intersected the white ice-seams, I found that the
latter penetrated the glacier only to a limited
depth, having the appearance of a kind of glacial
'trap' intruded from above.

But how is the backward dip of the blue veins to be accounted for ?   Doubtless in the following way : At the base of the cascade the glacier is forcibly compressed by the thrust of the mass behind it ; besides this, it changes its inclination suddenly and considerably ; it is bent upwards, and the consequence of this bending is a system of wrinkles, such as those represented in the next figure.   The interior of a bent umbrella-handle sometimes presents wrinkles which are the representatives, in little, of the protuberances upon the glacier.   The coat-sleeve is an equally instructive illustration : when the arm

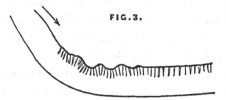

FIG. 3.

is bent at the elbow the sleeve wrinkles, and as the places where these wrinkles occur in the cloth are determined, to some extent, by the previous creasing, so also the places where the wrinkles are formed upon the glacier are determined by the previous scarring of the ice during its descent down the cascade.   The manner in which these crumples tend to *scale off* speaks strongly in favour of the explanation given.   The following figure represents a

type of numerous instances of scaling off.   By means
of a hydraulic press it is easy to produce a perfectly
similar scaling in small masses of ice.   One consequence of this crumpling of the glacier would be
the backward and forward inclination of the veins

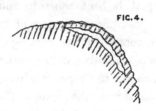

FIG. 4.

as actually observed.   The same appearance was
noticed on the wrinkles of the Glacier du Géant.
It was also proved, by measurements, that these
wrinkles *shorten* as they descend.

In virtue of what quality, then, can ice be bent and
squeezed, and have its form changed in the manner
indicated in the foregoing observations?   The only
theory worthy of serious consideration at the present day is the celebrated Viscous Theory of glacial
motion.   Numerous appearances, as we have seen,
favour the idea that ice is a viscous or 'semi-fluid'
substance, and that it flows as such in the glaciers
of the Alps.   The aspect of many glaciers, as a whole
—their power of closing up crevasses, and of recon-
structing themselves after having been precipitated
down glacial gorges—the obvious bendings and

contortions of various portions of the ice, are all in
harmony with the notion. The laminar structure
of the glacier has also been regarded by eminent
authorities as a crucial test in favour of the viscous
theory, and affirmed to be impossible of explanation
on any other hypothesis.

Nevertheless, this theory is so directly opposed to
our ordinary experience of the nature of ice as to
leave upon the mind a lingering doubt of its truth.
Can we imitate the phenomena without invoking the
explanation? We can. Moulds of various forms
were hollowed out in boxwood, and pieces of ice
were placed in these moulds and subjected to pressure. In this way spheres of ice were flattened into
cakes, and cakes formed into transparent lenses. A
straight bar of ice, six inches long, was passed
through a series of moulds augmenting in curvature,
and was finally bent into a semi-ring. A small
block of ice was placed in a hemispherical cavity,
and was pressed upon by a hemispherical protuberance, not large enough to fill the cavity; the ice
yielded and filled the space between both, thus
forming itself into a transparent cup. The specimens of ice here employed were so exceedingly
brittle that a pricker driven into the ice was competent to split blocks of the substance eight cubic
feet in volume, the surface of fracture being in all
cases as clean and sharp as that of glass.

These experiments, then, demonstrate a capacity on the part of small masses of ice which they have not been hitherto known to possess. They prove, to all appearance, that the substance is much more plastic than it was ever imagined to be. But the real germ from which these results have sprung is to be found in a lecture given at the Royal Institution in June 1850, and reported in the 'Athenæum' and 'Literary Gazette' for that year. Faraday then showed that when two pieces of ice, at a temperature of 32° Fahr., are placed in contact with each other, they freeze together, by the conversion of the film of moisture between them into ice. The case of a snowball is a familiar illustration of the principle. When the snow is below 32°, and therefore *dry*, it will not cohere, whereas when it is in a thawing condition it can be squeezed into a hard mass. During one of the hottest days of July 1857, when the thermometer was upwards of 100° Fahr. in the sun, and more than 80° in the shade, I observed a number of blocks of ice, which had been placed in a heap, frozen together at their places of contact; and I afterwards caused them to freeze together under water as hot as the hand could bear. Facts like these suggested the thought that if a piece of ice—a straight prism, for example—were placed in a bent mould and subjected to pressure it would break, but that the force would also bring its

ruptured surfaces into contact, and thus the continuity of the mass might be re-established. Experiment, as we have seen, completely confirmed
this surmise : the ice passed from a continuous
straight bar to a continuous bent one, the transition
being effected, not by a viscous movement of the
particles, but *through fracture and regelation.*

Let the transition from curve to curve be only
gradual enough, and we have the exact case of a
transverse slice of a glacier.

All the phenomena of motion, on which the idea of
viscosity has been based, are brought by such experiments as the above into harmony with the demonstrable properties of ice. In virtue of this property,
the glacier accommodates itself to its bed while preserving its general continuity, crevasses are closed
up, and the broken ice of a cascade, such as that of
the Talèfre or the Rhone, is recompacted to a solid
continuous mass.

The very essence of viscosity is the ability of
yielding to a force of tension, *the texture of the substance, after yielding, being in a state of equilibrium,*
so that it has no strain to recover from; and the
substances chosen by Prof. Forbes as illustrative
of the physical condition of a glacier possess this
power of being drawn out in a very eminent degree.
But it has been urged, and justly urged, that we

ought not to conclude that viscosity is absent because hand specimens are brittle, any more than we ought to conclude that ice is not blue because small fragments of the substance do not exhibit this colour. To test the question of viscosity, then, we must appeal to the glacier itself. Let us do so.

An analogy between the motion of a glacier through a sinuous valley and of a river in a sinuous channel has been already pointed out. But the analogy fails in one important particular: the river, and much more so a mass of flowing treacle, honey, tar, or melted caoutchouc, sweeps round its curves without rupture of continuity. The viscous mass *stretches*, but the icy mass *breaks*, and the 'excessive crevassing' pointed out by Prof. Forbes himself is the consequence. The inclinations of the Mer de Glace and its three tributaries were, moreover, taken, and the association of transverse crevasses with the changes of inclination were accurately noted. Every traveller knows the utter dislocation and confusion produced by the descent of the Mer de Glace from the Chapeau downwards. A similar state of things exists in the ice-cascade of the Talèfre. Descending from the Jardin, as the ice approaches the fall, great transverse chasms are formed, which at length follow each other so speedily as to reduce the ice-masses between them to mere plates and wedges, along which the explorer has to creep cautiously. These plates and wedges are in some cases bent and crumpled by the lateral pressure, and some large pyramids are turned 90° round, so as to have their veins at right angles to the normal position. The ice afterwards descends the fall, the portions exposed to view being a fantastic assemblage of frozen boulders, pinnacles, and towers, some erect, some leaning, falling at intervals with a sound like thunder, and crushing the ice-crags on which they fall to powder. The descent of the ice through this fall has been referred to as a proof of its viscosity: but the description just given does not harmonise with our ideas of a viscous substance.

But the proof of the non-viscosity of the substance must be sought at places where the change of inclination is very small. Nearly opposite l'Angle there is a change from four to nine degrees, and the consequence is the production of transverse fissures which render the glacier here perfectly impassable. Further up the glacier transverse crevasses are produced by a change of inclination from three to five degrees. This change of inclination is protracted

Fig. 5.

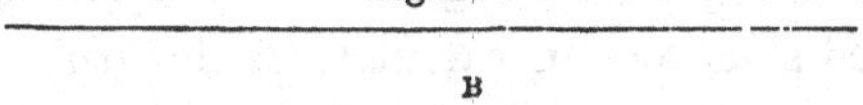

B

in fig. 5; the bend occurs at the point B; it is scarcely perceptible, and still the glacier is unable to pass over it without breaking across.

Again, the crevasses being due to a state of strain from which the ice relieves itself by breaking, the rate at which they widen may be taken as a measure of the amount of relief demanded by the ice. Both the suddenness of their formation and the slowness with which they widen are demonstrative of the non-viscosity of the ice. For were the substance capable of stretching, even at the small rate at which they widen, there would be no necessity for their formation.

Further, the marginal crevasses of a glacier are known to be a consequence of the swifter flow of its central portions, which throws the sides into a state of strain from which they relieve themselves by breaking. Now it is easy to calculate the amount of stretching demanded of the ice in order to accommodate itself to the speedier central flow. Take the case of a glacier half a mile wide. A straight transverse element, or slice, of such a glacier, is bent in twenty-four hours to a curve. The ends of the slice move a little, but the centre moves more: let us suppose the versed side of the curve formed by the slice in twenty-four hours to be a foot, which is a fair average. Having the chord of this arc, and its versed side, we can calculate its length. In the case of the Mer de Glace, which is about half a mile wide, the amount of stretching demanded would be about the eightieth of an inch in twenty-four hours. Surely, if the glacier possessed a property which could with any propriety be called viscosity, it ought to be able to respond to this

moderate demand; but it is not able to do so:
instead of stretching as a viscous body, in obedience
to this slow strain, it breaks as an eminently fragile
one, and marginal crevasses are the consequence.
It may be urged that it is not fair to distribute the
strain over the entire length of the curve: but reduce the distance as we may, a residue must remain,
which is demonstrative of the non-viscosity of the ice.

To sum up, then, two classes of facts present
themselves to the glacier investigator—one class in
harmony with the idea of viscosity, and another as
distinctly opposed to it. Where *pressure* comes
into play we have the former; where *tension* comes
into play we have the latter. Both classes of facts
are reconciled by the assumption, or rather the
experimental verity, that the fragility of ice
and its power of regelation render it possible for
it to change its form without prejudice to its
continuity.

[Very interesting experiments upon the bending of
ice have been recently made by Mr. Matthews and
Mr. Froude. In these experiments the temperature
of the ice, I believe, was some degrees below the
freezing point: it would be important to repeat
these experiments with ice at the temperature
which it actually possesses in glaciers, namely, at
32°.—April 1871.]

II.

STRUCTURE AND PROPERTIES OF ICE.

BEING desirous of examining how the interior of a mass of ice is affected by a beam of radiant heat sent through it, I availed myself of the sunny weather of September and October 1857. The sunbeams, condensed by a lens, were sent in various directions through slabs of ice. The path of every beam was observed to be instantly studded with lustrous spots, which increased in magnitude and number as the action continued. On examining the spots more closely, they were found to be flattened spheroids, and around each of them the ice was so liquefied as to form a beautiful flower-shaped figure possessing six petals. From this number there was no deviation. At first the edges of the liquid leaves were unindented; but a continuance of the action usually caused the edges to become serrated like those of ferns. When the ice was caused to move across the beam, or when the beam was caused to traverse different portions of the ice in succession, the sudden generation and crowding together of these liquid flowers, with their central spots shining with more than metallic brilliancy, was exceedingly beautiful.

In almost all cases the flowers were formed in planes parallel to the surface of freezing ; it mattered not whether the beam traversed the ice parallel to this surface or perpendicular to it. Some apparent exceptions to this rule were found, which will form the subject of future investigation.

The general appearance of the shining spots at the centres of the flowers was that of the bubbles of air entrapped in the ice ; to examine whether they contained air or not, portions of ice containing them were immersed in warm water. When the ice surrounding the cavities had completely melted, the latter instantly collapsed, and no trace of air rose to the surface of the water. A vacuum, therefore, had been formed at the centre of each spot, due, doubtless, to the well-known fact that the volume of water in each flower was less than that of the ice, by the melting of which the flower was produced.

The associated air-and-water cells, found in such numbers in the ice of glaciers, and also observed in lake ice, were next examined. Two hypotheses have been started to account for these cells. One attributes them to the absorption of the sun's heat by the air of the bubbles, and the consequent melting of the ice which surrounds them. The other hypothesis supposes that the liquid in the cells never has been frozen, but has continued in the liquid condition from the *névé* or origin of the glacier downwards. Now if the water in the cells be due to the melting of the ice, the associated air must be *rarefied*, because the volume of the liquid is less than that of the ice which produced it; whereas if the air be simply that entrapped in the snow of the *névé*, it will not be thus rarefied. Here, then, we have a test as to whether the water-cells have been produced by the melting of the ice.

Portions of ice containing these compound cells were immersed in hot water, the ice around the cavities being thus gradually melted away. When a liquid connexion was established between the bubble and the atmosphere, the former collapsed to a smaller bubble. In many cases the residual bubble did not reach the hundredth part of the magnitude of the primitive one. There was no exception to this rule, and it proves that the water of these particular cavities, at all events, is really due to the melting of the adjacent ice.

But how was the ice surrounding the bubbles melted? The hypothesis that the melting is due to the absorption of the solar rays by the air of the bubbles is that of M. Agassiz, which has been re-produced and subscribed to by the Messrs. Schla-gintweit, and accepted generally as the true one. Let us pursue it to its consequences.

Comparing equal *weights* of air and water, experiment proves that to raise a given weight of water one degree in temperature, as much heat would be needed as would raise the same weight of air four degrees.

Comparing equal *volumes* of air and water, the water is known to be 770 times heavier than the air; consequently, for a given volume of air to raise an equal volume of water one degree in temperature, it must part with $770 \times 4 = 3080$ degrees.

Now the quantity of heat necessary to melt a given weight of ice would raise the same weight of water 142·6 degrees Fahr. in temperature. Hence to produce, by the melting of ice, an amount of water equal to itself in bulk, a bubble of air must yield up $3080 \times 142·6$, or upwards of four hundred thousand degrees Fahrenheit.

This is the amount of heat which, according to the hypothesis of M. Agassiz and the Messrs. Schlagintweit, is absorbed by the bubble of air under the eyes of the observer. That is to say, the air is capable of absorbing an amount of heat which, had it not been communicated to the surrounding ice, would raise the bubble to a temperature 160 times that of fused cast iron. Did air possess this enormous power of absorption it would not be without inconvenience for the animal and vegetable life of our planet.

The fact is, that a bubble of air at the earth's surface is unable, in the slightest appreciable degree, to absorb the sun's rays; for those rays before they reach the earth have been perfectly sifted by their passage through the atmosphere. I made the following experiment illustrative of this point: The rays from an electric lamp were condensed by a lens, and the concentrated beam sent through the bulb of a differential thermometer. The heat of the beam was intense; still not the slightest effect was produced upon the thermometer. In fact, all the rays that *air* could absorb had been absorbed before the thermometer was reached, while the rays that glass could absorb had been *absorbed by the lens.* The heat consequently passed through the thin glass envelope of the thermometer, and the air within it, without imparting the slightest sensible heat to either.

The liquid bubbles observed in lake ice, and those which occur in the deeper portions of glacier ice, are produced by heat which has been *conducted* through the substance without melting it. Regarding heat as a mode of motion, it seems natural to infer, that inasmuch as within the mass each molecule is controlled in its motion by the surrounding molecules, the liberty of liquidity must be attained by the molecules at the surface of ice before the molecules in the interior can attain this liberty. But if a

cavity exist in the interior, the molecules surrounding that cavity are in a condition similar to those
at the surface; and they may be liberated by an
amount of motion which has been transmitted
through the ice without prejudice to its solidity.
The conception is helped when we call to mind the
transmission of motion through a series of elastic
balls, by which the last ball of the series is detached,
while the others do not suffer visible separation.
It may indeed be proved, by actual experiment,
that the interior portion of a mass of ice can be
liquefied by an amount of heat which has been
conducted through the exterior portions without
melting them.

Now precisely the converse of this takes place
when two pieces of ice, at 32° Fahr., with moist
surfaces, are brought into contact.   Superficial
portions are by this act transferred to the centre,
where a temperature of 32° is not quite sufficient
to produce liquefaction.   The motion of liquidity
which the surfaces possessed before contact is now
checked, and the pieces of ice freeze together.   This
appears to furnish a satisfactory explanation of all
the cases of this nature which have hitherto been
observed.

The particles of a crushed mass of ice at 32°, or a
ball of moist snow, may, it is now well known, be
squeezed into slabs or cups of ice.   That moisture is

necessary here, and that the same agent is necessary
in the conversion of snow into glacier ice was proved
by the following experiment:—A ball of ice was
cooled in a bath of solid carbonic acid and ether,
and thus rendered perfectly dry. Placed in a suitable mould, and subjected to hydraulic pressure,
the ball was crushed; but the crushed fragments
remained as *white and opaque* as those of crushed
glass. The particles, while thus dry, could not be
squeezed so as to form pellucid ice, which is so
easily obtained when the compressed mass is at a
temperature of 32° Fahr.

III.

STRUCTURE OF GLACIERS.

IF a transparent colourless solid be reduced to powder, the powder is white. Thus rock crystal, rock salt, and glass in powder, are all white. A glass jar, partially filled with a solution of carbonate of soda, with a little gum added to give it tenacity, presents, on the addition of a little tartaric acid, the appearance of a tall white column of foam. In all these cases, the whiteness and the opacity are due to the intimate and irregular admixture of a solid or a liquid with air ; in like manner the whiteness of snow is due to the mixture of air and transparent particles of ice.

The snow falls upon mountain eminences, and, above the snow-line, each year leaves a residue ; the substance thus collects in layers, forming masses of great depth. The lower portions are squeezed by the pressure of those above them, and a gradual approach to ice is the consequence. The air being gradually expelled, the transparency of the substance augments in proportion.

But even after the snow has been squeezed to hard ice in the upper glacier region, it always contains a large amount of the air originally entrapped in the snow. The air is distributed through the solid in the form of bubbles, which give the ice a milky appearance. At the lower extremity of a glacier the ice, as everybody knows, is blue and transparent. The transition from one state to the other is not, in all cases, a gradual change which takes place uniformly throughout the entire mass. The white ice, on the contrary, of the middle glacier region is usually striped by veins of a more transparent character, the air which gives to the ice its whiteness having been, by some means or other, wholly or partially ejected from the veins. These veins sometimes give the ice of many glaciers a beautiful laminated appearance; vast portions, indeed, of various glaciers consist of this laminated ice.

The theory of the veins which perhaps first presents itself to the mind, and which is still entertained by many intelligent Alpine explorers, is that the veining of the middle glaciers is simply a continuation of the *bedding* of the *névé*; that not only do the annual snow-falls produce beds of great thickness, but every successive fall tends to produce a layer of less thickness, which layers, or the surfaces separating them, ultimately appear as the blue veins.

This theory demands respectful consideration : on the exposed sections of the *névé* the lines of stratification are very manifest, exhibiting in many cases appearances strongly resembling that of the veined structure. Indeed, it was with a view to examine this subject more closely that I withheld my observations on the structure of the Mer de Glace in 1857, and betook myself once more to the mountains during the summer of 1858. My desire at that time was to settle once for all the rival claims of the only two theories which then deserved serious attention—namely, those of pressure and of stratification.

In pursuance of this idea, I first visited the Lower glacier of Grindelwald, one of the most accessible, and at the same time most instructive, in the entire range of the Alps. Ascending the branch of this glacier which descends from the Schreckhorn, the Strahleck, and the Finsteraarhorn, I came to the base of an ice-fall which forbade further advance. Quitting the glacier here, I ascended the side of the flanking mountain, so as to reach a point from which the fall, and the glacier below it, are distinctly visible ; and from this position I observed the gradual development and perfecting of the structure at the base of the fall. On the middle of the fall itself no trace of the structure was manifest ; but where the glacier changed its inclination at the bottom, being bent upwards so as throw its surface into a state of intense longitudinal compression, the blue veins first made their appearance. The base of the fall was a true *structure mill*, where the transverse veins were manufactured, being afterwards sent forward, giving a character to portions of the glacier which had no share in their formation.

I afterwards examined the fall from the opposite side of the valley, and corroborated the observations. It is difficult, in words, to convey the force of the evidence which this glacier presents to the observer who *sees* it ; it seems in fact like a grand laboratory experiment made by Nature herself with especial reference to the point in question. The squeezing of the mass, its yielding to the force brought to bear upon it, its wrinkling and scaling off, and the appearance of the veins at the exact point where the pressure begins to manifest itself, leave no doubt on the mind that pressure and structure stand to each other in the relation of cause and effect, and that the stratification could have nothing to do with the phenomenon.

I subsequently crossed the Strahleck, descended the glaciers of the Aar, crossed the Grimsel, and examined the glacier of the Rhone. This glacier has also its grand ice-fall. In company with Prof. Ramsay, I climbed in 1858 the precipices flanking the fall at the Grimsel side. What has been stated regarding the Grindelwald ice-fall is true of that of the Rhone; the base of the cascade is *the manufactory of the structure*; and, as all the ice has to pass through this mill, the entire mass of the glacier from the base of the fall downwards is beautifully laminated.

Descending the valley of the Rhone to Viesch, I went thence to the Æggischhorn, and remained for eight days in the vicinity of the Great Aletsch glacier — the noblest ice-stream of the Alps. A highly intelligent explorer had adduced certain phenomena of this glacier as an evidence against the pressure theory of the veined structure; and I did not think myself justified in quitting the place until I had perfectly satisfied myself that the Aletsch not only presented no phenomena at variance with the pressure theory, but exhibited some which seemed fatal to the theory of the stratification.

I subsequently proceeded to Zermatt, and spent ten days on the Riffelberg, exploring the entire system of glaciers between Monte Rosa and the Mont Cervin. These glaciers exhibit, perhaps in a more striking manner than any others in the Alps, the yielding of glacier ice when subjected to intense pressure. The great western glacier of Monte Rosa, the Schwartze glacier, the Trifti glacier, and the glaciers of St. Theodule, are first spread out as wide and extensive *névés* over the breasts of the mountains.

They move down, and are finally forced into the valley containing the trunk, or Görner glacier. Here they are squeezed to narrow strips, which gradually dwindle in width until they form driblets not more than a few yards across. From the Görner Grat, or from the summit of the Riffelhorn, these parallel strips of glacier, each separated from its neighbour by a medial moraine, present a most striking and instructive appearance.

The structure of these glaciers was carefully examined, and in all cases as I travelled from regions where the pressure was feeble to others where it was intense, the ice changed from a state almost, if not entirely, structureless, to one in which the veining was exhibited in great perfection. Each glacier, for example, where it met the opposing mass in the trunk valley, and was pressed against the latter by the thrust from behind, exhibited a beautifully developed structure.

Proofs have been already adduced that the Glacier du Géant is in a state of longitudinal compression ; it has also been shown that the seams of white ice which intersect this glacier are due to the filling up of the channels of glacier streams by snow, and the subsequent compression of the substance. Here, then, we have a vast ice-press which furnishes us with a test of the pressure theory. Both in 1857 and 1858 I found many of these seams of white ice intersected by blue veins of the finest and most distinct character, their general direction being at right angles to the direction of pressure.

But the notions of M. Agassiz as to the turning up of the strata so as to expose their edges at the surface, and the acute remarks and arguments of Mr. John Ball on the same subject, might still cast a doubt upon the pressure theory, by suggesting a possible, though extremely improbable, explanation of the structure in accordance with the theory of stratification.

Hence my strong desire to discover some crucial phenomenon which should set this question for ever at rest, and leave no room for doubt, even on the minds of those who never saw a glacier. On Wednesday, August 18, I was fortunate enough to make this discovery upon the Furgge glacier.

This ice-field spreads out as an almost level plain at the base of Mont Cervin. The strata pile themselves one above the other without disturbance, and hence with great regularity. The ice at length reaches a brow, over which it is precipitated, forming in its descent four great terraces, and shutting up the lower valley as a *cul de sac*. When I reached this place huge blocks of ice stood, like rocking stones, upon the topmost ledge, and numbers, which had fallen, had been caught by the other ledges, and occupied very threatening positions: the base of the fall was cumbered with crushed ice, and large boulders of the substance had been cast a considerable way down the glacier.

On the faces of the terraces horizontal lines of stratification were shown in the most perfect manner. Here and there the exertion of a powerful lateral squeeze was manifest, causing the beds to crumple, and producing numerous faults. Examining the fall from a distance through an opera-glass, I thought I could discover lines of veining running *through* the strata, at a high angle, exactly as the planes of cleavage often run at a high angle to the bedding of slate rocks. The surface of the ice was, however, weathered; and I was unwilling to accept an observation upon such a cardinal point with a shade of doubt attached to it. Leaving my field-glass with my guide, who was to give me warning should the blocks overhead give way, I advanced to the wall of ice, and at several places cut away with my axe the weathered superficial portions. Underneath I found the true veined structure, *running nearly at right angles to the planes of stratification.*

I afterwards climbed the glacier to the right, and, as I ascended, still better illustrations of the co-existence of the structure and the strata than those observed upon the terraces exhibited themselves. The ice was greatly dislocated, and on the faces of the crevasses the beds were distinctly shown, *with the veins crossing them.* The idea that the veins could be due to the turning up of the strata is plainly irreconcileable with these observations.

The same year I visited the Mer de Glace and its tributaries, and found the pressure key applicable to their phenomena also. The transverse structure of the Glacier du Géant is formed at the base of the séracs; that of the Talèfre branch of the Mer de Glace at the base of the Talèfre ice-fall, where the change of inclination and the thrust from behind produce the requisite longitudinal compression. I have already had occasion to remark upon the peculiar dipping of the structure, and the scaling-off of the protuberances, which are effects of the same cause. These phenomena are exhibited at the base of all the ice-cascades.

The principal kinds of structure may be divided into three; as follows:

1st, *Marginal structure,* developed by pressure due to the swifter motion of the centre of the glacier.

2nd, *Longitudinal structure,* due to mutual pressure of two tributary glaciers; the structure here is parallel to the medial moraine which divides the tributaries.

3rd, *Transverse structure,* produced by pressure

due to the change of inclination, and to the longitudinal thrust endured by the glacier at the base
of an ice-fall.

The lamination of a glacier is a peculiarly interesting case of cleavage. It is produced in the same
manner as the lamination of slate rock, which is
known, through the distortion of its fossils, to have
suffered great pressure at right angles to the planes
of cleavage.

IV.

HELMHOLTZ ON ICE AND GLACIERS.

SWITZERLAND has attractions for the scientific
philosophers of Germany, and around the Titlis,
Bunsen, Helmholtz, Kirchhoff, and Wiedemann are
not unfamiliar names. Nor have their visits to the
Alps been unproductive of results. Some time ago
I was favoured by Professor Helmholtz with the
First Part of his 'Popular Scientific Lectures.' It
contains four of them—the first, 'On the Relation
of the Natural Sciences to Science in general;' the
second, 'On Goethe's Labours in Natural Science;'
the third, 'On the Physiological Origin of Musical
Harmony;' and the fourth, 'On Ice and Glaciers.'
The lectures are in German, and it is much to be
desired that some competent person should undertake their translation into English.[1]

I turned with natural interest to the last-men-
tioned discourse, to see how my notions and experiments on the formation and motion of glaciers were

[1] I have reason to believe that a translation of the two parts
hitherto published will soon be forthcoming.—J. T., 1871.

regarded by so eminent a man. I will here endeavour to give a summary of the scientific portion
of the lecture.

Professor Helmholtz refers the cold of the upper
regions of the atmosphere to the causes generally
assigned; but he adds a remark important at the
present moment, when the origin of the hot wind
called Föhn in Switzerland is the subject of so
much discussion. This wind, as Helmholtz justly
observes, may not only be a cold wind upon the
mountain-summits, but a *wet* one, and it may
deposit its moisture there. A wind thus dried upon
the heights, and warmed by its subsequent fall into
the valleys, would possess the heat and dryness of
the Föhn. These qualities are, therefore, no proof
that the origin of the Föhnwind is Sahara.

It will probably be remembered that I deduced
the formation of glaciers, and their subsequent
motion through valleys of varying width and flexure,
from the fact that when two pieces of ice are pressed
together they freeze together at their places of
contact. This fact was first mentioned to me
verbally by its discoverer, Faraday. Soon afterwards, and long before I had occasion to reflect
upon its cause, the application of the fact to the
formation and motion of glaciers flashed upon me.
Snow was in the yard of the Royal Institution at the
time; stuffing a quantity of it into a steel mould,

which I had previously employed to demonstrate
the influence of pressure on magnetic phenomena,
I squeezed the snow, and had the pleasure of seeing
it turn out from the mould as a cylinder of trans-

Fig. 6.

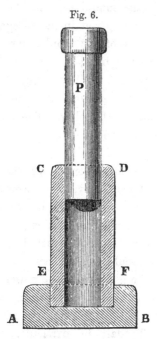

lucent ice. I immediately went to Faraday, and
expressed the conviction that his little outlying
experiment would be found to constitute the basis
of a true theory of glaciers. It became subsequently
known to me that the Messrs. Schlagintweit had

made a similar experiment with snow; but they did
not connect with it the applications which suggested
themselves to me, and which have since been developed into a theory of glacier-motion.

A section of the mould used in the experiment
above referred to is given in the foregoing figure.
A B is the solid base of the mould; C D E F a hollow
cylinder let into the base; P is the solid plug used
to compress the snow. When sufficiently squeezed,
the bottom, A B, is removed, and the cylinder of
ice is pushed out by the plug. The mould closely
resembles one of those employed by Professor
Helmholtz.

The subsequent development of the subject by
the moulding of ice into various forms by pressure
is too well known to need dwelling upon here. In
applying these results to glaciers, I dwelt with
especial emphasis upon the fact that while the
power of being moulded by *pressure* belonged in an
eminent degree to glacier ice, the power of yielding,
by stretching, to a force of *tension*, was sensibly
wanting. On this point Prof. Helmholtz speaks
as follows: 'Tyndall in particular maintained,
and proved by calculation and measurement, that
the ice of a glacier does not stretch in the smallest
degree when subjected to tension—that when sufficiently strained it always breaks;' and he adds,
in another place, that the property thus revealed

establishes ' an essential difference between a stream of ice, and one of lava, tar, honey, or mud.'

In the beautiful experiments of M. Tresca recently executed, the power of ice to mould itself under pressure has been very strikingly illustrated. Professor Helmholtz also, in the presence of his audiences at Heidelberg and Frankfort, illustrated this property in various ways. From snow and

Fig. 7.   Fig. 8.

broken fragments of ice he formed cakes and cylinders; and uniting the latter, end to end, he permitted them to freeze together to long sticks of ice. Placing, moreover, in a suitable mould a cylinder of ice of the shape represented in fig. 7, he squeezed it into the cake represented in fig. 8. In fact he corroborated, by a series of striking experimental devices of his own the results previously obtained by myself.

With regard to the application of these results to the phenomena of glaciers, Professor Helmholtz, after satisfying himself of the insufficiency of other hypotheses, thus finally expresses his conviction: 'I do not doubt that Tyndall has assigned the essential and principal cause of glacier-motion, in referring it to fracture and regelation.'

It is perhaps worth stating that the term 'regelation' was first introduced in a paper published by Mr. Huxley and myself more than seven years after the discovery of the fact by Faraday, and that it was suggested to us by our friend Dr. Hooker, Director of the Royal Gardens at Kew. As already remarked, the formation and motion of glaciers, and other points of a kindred nature, had been referred to regelation long before I occupied myself with the cause of regelation itself. This latter question is not once referred to in the memoir in which the regelation theory was first developed.[1] The enquiries, though related, were different. In referring the motion of glaciers to a fact experimentally demonstrated, I referred it to its *proximate* cause. To refer that cause to *its* physical antecedents formed the subject of a distinct enquiry, in which, because of my belief in the substantial correctness of Faraday's explanation, I took comparatively little part.

[1] Phil. Trans. vol. cxlvii. p. 327.

Five persons, however, mingled more or less in the enquiry—viz. Professor Faraday, Principal Forbes, Professor James Thomson, Professor (now Sir) William Thomson, and myself.[1] Professor James Thomson explained regelation by reference to an important deduction, first drawn by him,[2] and almost simultaneously by Professor Clausius,[3] from the mechanical theory of heat. He had shown it to be a consequence of this theory that the freezing-point of water must be lowered by pressure; that is to say, water when subjected to pressure will remain liquid at a temperature below that at which it would freeze if the pressure were removed. This theoretic deduction was confirmed in a remarkable manner by the experiments of his brother.[4] Regelation, according to James Thomson's theory, was thus accounted for: 'When two pieces of ice are pressed together, or laid the one upon the other, their compressed parts liquefy. The water thus produced has rendered latent a portion of the heat of the surrounding ice, and must therefore be lower than 0° C. in temperature. On escaping from the pressure this water refreezes and cements the pieces of ice together.

[1] Proc. Roy. Soc. vol. ix. p. 141 ; and vol. x. p. 152. Phil. Mag. S. 4, vol. xvi. pp. 347 and 544 ; and vol. xvii. p. 162.

[2] Proc. Roy. Soc. Edinb. February 1850.

[3] Pogg. Ann. vol. lxxxi. p. 168.

[4] Phil. Mag. August 1850.

I always admitted that this explanation dealt with a 'true cause.' But considering the infinitesimal magnitude of the pressure sufficient to produce re-gelation, in common with Professor Faraday and Principal Forbes, I deemed the cause an insufficient one. Professor James Thomson, moreover, grounded upon the foregoing theory of regelation a theory of glacier-motion, in which he ascribed the changes of form which a glacier undergoes to the incessant liquefaction of the ice at places where the pressure is intense, and the refreezing, in other positions, of the water thus produced.[1] I endeavoured to show that this theory was inapplicable to the facts. Professor Helmholtz has recently subjected it to the test of experiment, and the conclusions which he draws from his researches are substantially the same as mine.

Thus, then, as regards the incapacity of the ice on which my observations were made to stretch in obedience to tension, and its capacity to be moulded to any extent by pressure—as regards the essential difference between a glacier, and a stream of lava, honey, or tar—as regards the sufficiency of pressure and regelation to account for the formation of glaciers, and of fracture and regelation to account for their motion—as regards, finally, the insuf-ficiency of the theory which refers the motion to

[1] Proc. Roy. Soc. vol. viii. p. 455.

liquefaction by pressure, and refreezing, the views
of Professor Helmholtz and myself appear to be
identical.

But the case is different with regard to the cause
of regelation itself.   Here Professor Helmholtz, like
M. Jamin,[1] accepts the clear and definite explanation
of Professor James Thomson as the most satisfactory
that has been advanced ; and he supports this view
by an experiment so beautiful that it cannot fail to
give pleasure even to those against whose opinions
it is adduced.   But before passing to the experiment,
which is described in the Appendix to the lecture,
it will be well to give in the words of Professor
Helmholtz the views which he expresses in the body
of his discourse.

‘ You will now ask with surprise,’ he says, ‘ how it
is that ice, the most fragile and brittle of all known
solid substances, can flow in a glacier like a viscous
mass ; and you may perhaps be inclined to regard
this as one of the most unnatural and paradoxical
assertions that ever was made by a natural philosopher.   I will at once admit that the enquirers
themselves were in no small degree perplexed by the
results of their investigations.   But the facts were
there, and could not be dissipated by denial.   How
this kind of motion on the part of ice was possible
remained long an enigma—the more so as the known

[1] ‘ Traité de Physique,’ vol. ii. p. 105.

brittleness of ice also manifested itself in glaciers by
the formation of numerous fissures.   This, as Tyndall
rightly maintained, constituted an essential difference between the ice-stream, and a stream of lava,
tar, honey, or mud.

'The solution of this wonderful enigma was found
—as is often the case in natural science—in an apparently remote investigation on the nature of heat,
which forms one of the most important conquests of
modern physics, and which is known under the name
of the mechanical theory of heat.   Among a great
number of deductions as to the relations of the most
diverse natural forces to each other, the principles of
the mechanical theory of heat enable us to draw
certain conclusions regarding the dependence of the
freezing-point of water on the pressure to which the
ice and water are subjected.'

Professor Helmholtz then explains to his audience
what is meant by latent heat, and points out that,
through the circulation of water in the fissures and
capillaries of a glacier, its interior temperature must
remain constantly at the freezing-point.

'But,' he continues, 'the temperature of the
freezing-point of water can be altered by pressure.
This was first deduced by James Thomson, and
almost simultaneously by Clausius, from the mechanical theory of heat ; and by the same deductions
even the magnitude of the change may be predicted.

For the pressure of every additional atmosphere, the freezing-point sinks 0°·0075 C. The brother of the gentleman first named, William Thomson, the celebrated Glasgow physicist, verified experimentally the theoretic deduction by compressing a mixture of ice and water in a suitable vessel. The mixture became colder and colder as the pressure was augmented, and by the exact amount which the mechanical theory of heat required.

'If, then, by pressure a mixture of ice and water can be rendered colder without the actual abstraction of heat, this can only occur by the liquefaction of the ice and the rendering of heat latent. And this is the reason why pressure can alter the point of congelation. . . . .

'In the experiment of William Thomson just referred to ice and water were enclosed in a solid vessel from which nothing could escape. The case is somewhat different when, as in the case of a glacier, the water of the compressed ice can escape through fissures. In this case the ice is compressed, but not the water which escapes. The pressed ice will become colder by a quantity corresponding to the lowering of its freezing-point by the pressure. But the freezing-point of the uncompressed water is not lowered. Here, then, we have ice colder than 0° C. in contact with water at 0° C. The consequence is, that round the place of pressure the

water will freeze and form new ice, while, on the
other hand, a portion of the compressed ice continues
to be melted (während dafür ein Theil des gepressten
Eises fortschmilzt).

'This occurs, for instance, when two pieces of ice
are simply pressed together. By the water which
freezes at the points of contact they are firmly
united to a continuous mass. When the pressure is
considerable, and the chilling consequently great, the
union occurs quickly, but it may also be effected by
a very slight pressure if sufficient time be afforded.
Faraday, who discovered this phenomenon, named it
the *regelation of ice*.[1] Its explanation has given
rise to considerable controversy: I have laid that
explanation before you which I consider to be the
most satisfactory.'

In the Appendix, Professor Helmholtz returns to
the subject thus handled in the body of his discourse. 'The theory of the regelation of ice, he
observes, 'has given rise to a scientific discussion
between Faraday and Tyndall on the one hand, and
James and William Thomson on the other. In the
text of this lecture I have adopted the theory of the
latter, and have therefore to justify myself for so
doing.' He then analyses the reasonings on both sides,
points out the theoretic difficulties of Faraday's

[1] I have corrected this slight inadvertence. We owe the name
to Hooker.

explanation, shows what a small pressure can accomplish if only sufficient time be granted to it, draws attention to the fact that when one piece of ice is placed upon another the pressure is not distributed over the whole of the two appressed surfaces, but is concentrated on a few points of contact. He also holds, with Professor James Thomson, that in an experiment devised by Principal Forbes even the capillary attraction exerted between two plates of ice is sufficient, in due time, to produce regelation. To illustrate the slow action of the small differences of temperature which here come into play Professor Helmholtz made the following experiment, to which reference has been already made.

'A glass flask with a drawn-out neck was half filled with water, which was boiled until all the air above it was driven out. The flask was then hermetically sealed. When cooled, the flask was void of air, and the water within it freed from the pressure of the atmosphere. As the water thus prepared can be cooled considerably below 0° C. before the first ice is formed, while when ice is in the flask it freezes at 0° C. [why? J. T.], the flask was in the first instance placed in a freezing mixture until the water was changed into ice. It was afterwards permitted to melt slowly in a place the temperature of which was +2° C., until the half of it was liquefied.

'The flask thus half filled with water having a disk of ice swimming upon it was placed in a mixture of ice and water, being quite surrounded by the mixture. After an hour the disk within the flask was frozen to the glass. By shaking the flask the disk was liberated, but it froze again. This occurred as often as the shaking was repeated. The flask was permitted to remain for eight days in the mixture, which was preserved throughout at a temperature of 0° C. During this time a number of very regular and sharply defined ice-crystals were formed, and augmented very slowly in size. This is perhaps the best method of obtaining beautifully formed crystals of ice.

'While, therefore, the outer ice which had to support the pressure of the atmosphere slowly melted, the water within the flask, whose freezing-point, on account of a defect of pressure, was 0°·0075 C. higher, deposited crystals of ice. The heat abstracted from the water in this operation had, moreover, to pass through the glass of the flask, which, together with the small difference of temperature, explains the slowness of the freezing process.'

A single additional condition in connection with this beautiful experiment I should like to have seen fulfilled—namely, that the water in which the flask was immersed, as well as that within it, should

be purged of its air by boiling. It is just possible
that the point of congelation may not be entirely
independent of the presence of air in the water.

The revival of this subject by Professor Helmholtz
has caused me to make a few additional experiments on the moulding and regelation of ice. The
following illustrates both :—A quantity of snowy
powder was scraped from a block of clear ice and

Fig. 9.   Fig. 10.

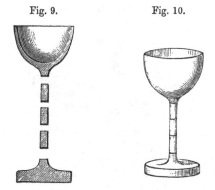

placed in a boxwood mould having a shape like
the foot of a claret-glass. The ice-powder being
squeezed by a hydraulic press, a clear mass of ice
of the shape shown in section at the bottom of
fig. 9 was the result. In another mould the same
powder was squeezed so as to form small cylinders,
three of which are shown separate in fig. 9. A third
mould was then employed to form a cup of ice,
which is shown at the top of fig. 9. Bringing all

the parts into contact, they were cemented through
regelation to form the claret-glass sketched in fig. 10,
from which several draughts of wine might be taken,
if the liquid were cooled sufficiently before pouring
it into the cup of ice.

There are brass shapes used for the casting of
flowers and other objects which answer admirably
for experiments on the regelation of ice.   One of

Fig. 11.

them was purchased for me by Mr. Becker. Ice-
powder squeezed into it regelated to a solid mass
and came from the mould in the sharply defined
form sketched in fig. 11.

I placed a small piece of ice in warm water and
pressed it underneath the water by a second piece.

The submerged morsel was so small that the vertical pressure was almost infinitesimal. It froze, notwithstanding, to the under surface of the superior piece of ice. Two pieces of ice were placed in a basin of warm water, and allowed to come together. They froze as soon as they touched each other. The parts surrounding the place of contact rapidly melted away, but the two pieces continued for a time united by a narrow bridge of ice. The bridge finally melted away, and the pieces were for a moment separated. But bodies which water wets, and against which it rises by capillary attraction, move spontaneously together upon water. The ice morsels did so, and immediately regelation again set in. A new bridge was formed, which in its turn was dissolved, and the pieces closed up as before. Thus a kind of pulsation was kept up by the two pieces of ice. They touched, froze, a bridge was formed and melted, leaving an interval between the pieces. Across this they moved, touched, froze, the same process being repeated over and over again.

We have here the explanation of the curious fact that when several large lumps of ice are placed in warm water and allowed to touch each other, regelation is maintained among them as long as they remain undissolved. The final fragments may not be the one-hundredth part of the original ones in size; but through the process just described, they incessantly lock themselves together until they finally disappear.

According to Professor James Thomson's theory, to produce regelation the pieces of ice have to exercise pressure, in order to draw from the surrounding ice the heat necessary for the liquefaction of the compressed part; and then this water must escape and be refrozen. All this requires time. In the foregoing experiments, moreover, the water liquefied by the pressure issued into the surrounding warm water, but notwithstanding this the floating fragments regelated in a moment. It is not necessary that the touching surfaces should be flat; for in this case a film of water might be supposed to exist between them of the temperature 0° C. The surfaces in contact may be convex: they may be virtual *points* that are about to touch each other, clasped all round by the warm liquid, which is rapidly dissolving them as they approach. Still they freeze immediately when they touch.

There are two points urged by Helmholtz—one in favour of the view he has adopted, and the other, showing a difficulty associated with the view of Faraday—on which a few words may be said. 'I found,' says Helmholtz, 'the strength and rapidity of the union of the pieces of ice in such complete correspondence with the amount of pressure employed, that I cannot doubt that the pressure is actually the sufficient cause of the union.'

But, according to Faraday's explanation, the strength and quickness of the regelation must also go hand in hand with the magnitude of the pressure employed. Helmholtz rightly dwells upon the fact that the appressed surfaces are usually not perfectly congruent—that they really touch each other in a few points only, the pressure being, therefore, concentrated. Now the effect of pressure exerted on two pieces of ice at a temperature of 0° C. is not only to lessen the thickness of the liquid film between the pieces, but also to flatten out the appressed points, and thus to spread the film over a greater space. On both theories, therefore, the strength and quickness of the regelation ought to correspond to the magnitude of the pressure.

The difficulty referred to above is thus stated by Helmholtz: 'In the explanation given by Faraday, according to which the regelation is caused by a contact action of ice and water, I find a theoretic difficulty. By the freezing of the water a very sensible quantity of heat would be set free; and it does not appear how this is to be disposed of.'

On the part of those who accept Faraday's explanation, the answer here would be that the free heat is diffused through the adjacent ice. But against this it will doubtless be urged that ice already at a temperature of 0° C. cannot take up more heat without liquefaction. If this be true under all circumstances, Faraday's explanation must undoubtedly be given up. But the essence of that explanation seems to be that the interior portions of a mass of ice require a higher temperature to dissolve them than that sufficient to cause fusion at the surface. When therefore two moist surfaces of ice at the temperature 0° are pressed together, and when, in virtue of the contact action assumed by Faraday, the film of water between them is frozen, the adjacent ice (which is now in the interior, and not at the surface as at first) is in a condition to withdraw by conduction, and without prejudice to its own solidity, the small amount of heat set free. Once granting the contact action claimed by Faraday, there seems to be no difficulty in disposing of the heat rendered sensible by the freezing of the film.

When the year is advanced, and after the ice imported into London has remained a long time in store, if closely examined, parcels of liquid water will be found in the interior of the mass. I enveloped ice containing such water-parcels in tinfoil, and placed it in a freezing mixture until the liquid parcels were perfectly congealed. Removing the ice from the freezing mixture, I placed it, covered by its envelope, in a dark room, and found, after a couple of hours' exposure to a temperature somewhat over 0° C., the frozen parcels again liquid. *The heat which fused this interior ice passed through the firmer surrounding ice without the slightest visible prejudice to its solidity.* But if the freezing temperature of the ice-parcels be 0° C., then the freezing temperature of the mass surrounding them must be higher than 0° C., which is what the explanation of Faraday requires.

In a quotation at p. 389 I have attached to the description of a precaution taken by Professor Helmholtz the query 'why?' He states that water freed of its air sinks, without freezing, to a temperature far below 0° C.; while when a piece of ice is in the water it cannot so sink in temperature, but is invariably deposited in the solid form at 0° C. This surely proves ice to possess a special power of solidification over water. It is needless to say that the fact is general—that a crystal of any salt placed in a saturated solution of the salt always provokes crystallisation. Applying this fact to the minute film of water enclosed between two appressed surfaces of ice, it seems to me in the highest degree probable that the contact action of Faraday will set in, that the film will freeze and cement the pieces of ice together.[1]

Apart from the present discussion, the following

---

[1] Both Professor Helmholtz and I have since agreed to consider the physical cause of regelation an open question.

observation is perhaps worth recording: It is well
known that ice during a thaw disintegrates so as to
form rude prisms whose axes are at right angles to
the planes of freezing. I have often observed this
action on a large scale during the winters that I
spent as a student on the banks of the Lahn. The
manner in which these prisms are in some cases
formed is extremely interesting. On close inspection, a kind of cloudiness is observed in the interior
of a mass of apparently perfect ice. Looked at
through a strong lens, this cloudiness appears as
striæ at right angles to the planes of freezing, and
when the direction of vision is across these planes
the ends of the striæ are apparent. The spaces between the striæ are composed of clear unclouded ice.
When duly magnified, the objects which produce the
striæ turn out to be piles of minute liquid flowers,
whose planes are at right angles to the direction of
the striæ.

Since writing the above, I have been favoured
with a copy of a discourse delivered by Professor
De la Rive, at the opening of the forty-ninth meeting of the Société Helvétique, which assembled in
1865 at Geneva. From this admirable *résumé* of
our present knowledge regarding glaciers I make
the following extract, which, together with those
from the lecture of Helmholtz, will show sufficiently

how the subject is now regarded by scientific men:
'Such, gentlemen,' says M. De la Rive, 'is a description of the phenomena of glaciers, and it now
remains to explain them, to consult observation,
and deduce from it the fundamental character of
the phenomena. Observation teaches us that gravity
is the motive force, and that this force acts upon a
solid body—ice—imparting to it a slow and continuous motion. What are we to conclude from this?
That ice is a solid which possesses the property of
flowing like a viscous body—a conclusion which
appears very simple, but which was nevertheless
announced for the first time hardly five-and-twenty
years ago by one of the most distinguished philosophers of Scotland, Professor James D. Forbes.
This theory, for it truly is a theory, basing itself on
facts as numerous as they are well observed, enun-
ciates the principle that ice possesses the characteristic properties which belong to plastic bodies.
Although he did not directly prove it, to Professor
Forbes belongs not the less the great merit of insist-
ing on the plasticity of ice, before Faraday, in dis-
covering the phenomenon of regelation, enabled
Tyndall to prove that the plasticity was real, at
least partially.

'The experiment of Faraday is classical in connexion with our subject. It consists, as you know,
in this, that if two morsels of ice be brought into

contact in water, which may be even warm, they freeze together. Tyndall immediately saw the application of Faraday's experiment to the theory of glaciers ; he comprehended that, since pieces of ice could thus solder themselves together, the substance might be broken, placed in a mould, compressed, and thus compelled to take the form of the cavity which contained it. A wooden mould, for example, embraces a spherical cavity; placing in it fragments of ice and squeezing them, we obtain an ice sphere ; placing this sphere in a second mould with a lenticular cavity and pressing it, we transform the sphere into a lens. In this way we can impart any form whatever to ice.

'Such is the discovery of Tyndall, which may well be thus named, particularly in view of its consequences. For all these moulds magnified become the borders of the valley in which a glacier flows. Here the action of the hydraulic press which has served for the experiments of the laboratory is replaced by the weight of the masses of snow and ice collected on the summits, and exerting their pressure on the ice which descends into the valley. Supposing, for example, between the spherical mould and the lenticular one, a graduated series of other moulds to exist, each of which differs very little from the one which precedes and from that which follows it, and that a mass of ice could be made to pass

through all these moulds in succession, the phenomenon would then become continuous. Instead of
rudely breaking, the ice would be compelled to change
by insensible degrees from the spherical to the
lenticular form. It would thus exhibit a plasticity
which might be compared to that of soft wax. But
ice is only plastic under *pressure*; it is not plastic
under *tension* : and this is the important point which
the vague theory of plasticity was unable to explain.
While a viscous body, like bitumen or honey, may
be drawn out in filaments by tension, ice, far from
stretching in this way, breaks like glass under this
action. These points well established by Tyndall, it
became easy for him to explain the mechanism of
glaciers, and by the aid of an English geometer, Mr.
William Hopkins, to show how the direction of the
crevasses of a glacier are the necessary consequences
of its motion.'

I have quite recently had a mould constructed for
me by Mr. Becker,[1] and yesterday (November 16,
1865) made with it an experiment which, on account
of the ease with which it may be performed, will interest all those who care about exhibiting in a striking and instructive manner the effects of regelation.

[1] I am continually indebted to this able mechanism for prompt
and intelligent aid in the carrying out of my ideas.

The mould is shown in fig. 12. It consists of two
pieces of cast iron, A B C and D F G, slightly wedge-
shaped and held together by the iron rectangle R E
which is slipped over them. The inner face of A B
C is shown in fig. 13. In it is hollowed out a semi-
ring M N, with a semicylindrical passage o leading

Fig. 12.

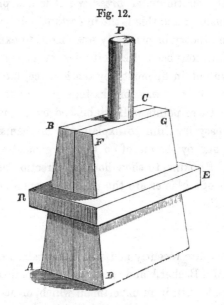

into it. The inner face of D F G is similarly hollowed out, so that when both faces are piaced
together, as in fig. 12, they enclose a ring 4 inches in
external diameter, from M to N, and ¾ of an inch in
thickness, with the passage o, 1 inch in diameter,

into which fits the polished iron plug P. At q and
r, fig. 13, are little pins which, fitting into holes
corresponding to them, keep the slabs A B C and D F G
from sliding over each other.

The mould being first cooled by placing it for a
short time in a mixture of ice and water, fragments
of ice are stuffed into the orifice O and driven down

Fig. 13.   Fig. 14.

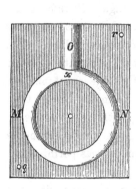
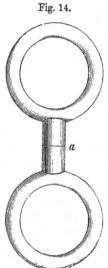

with a hammer by means of the plug P. The bruised
and broken ice separates at x, one portion going to
the right, the other to the left. Driving the ice
thus into the mould, piece after piece, it is finally
filled. By removing the rectangle R E, the two

halves of the mould are then separated, and a perfect ring of ice is found within. Two such rings soldered by regelation at a are shown in fig. 14. It would be easy thus to construct a chain of ice. An hydraulic press may of course be employed in this experiment, but it is not necessary; with the hammer and plug beautiful rings of ice are easily obtained by the regelation of the crushed fragments.

I have now to add the description of an experiment which suggested itself to my ingenious friend Mr. Duppa, when he saw the ice-rings just referred to, and which was actually executed by him yesterday (the 16th) in the laboratory of the Royal Institution. Pouring a quantity of plaster of paris into a proper vessel, an ice-ring was laid upon the substance, an additional quantity of the cement being then poured over the ring. The plaster ' set,' enclosing the ring within it: the ring soon melted, leaving its perfect matrix behind. The mould was permitted to dry, and, molten lead being poured into the space previously occupied by the ice, a leaden ring was produced. Now ice can be moulded into any shape : statuettes, vases, flowers, and innumerable other ornaments can be formed from it. These enclosed in cement, in the manner suggested by Mr. Duppa, remain intact sufficiently long to enable the cement to set around them; they afterwards melt and disappear, leaving behind them perfect plaster moulds, from which casts can be taken.

V.

CLOUDS.

FROM every natural fact invisible relations radiate, the apprehension of which imparts a measure of delight; and there is a store of pleasure of this kind ever at hand for those who have the capacity to turn natural appearances to account. It is pleasant, for example, to lie on one's back upon a dry green slope and watch the clouds forming and disappearing in the blue heaven. A few days back the firmament was mottled with floating cumuli, from the fringes of which light of dazzling whiteness was reflected downwards, while the chief mass of the clouds lay in dark shadow. From the edge of one large cloud-field stretched small streamers, which, when attentively observed, were seen to disappear gradually, and finally to leave no trace upon the blue sky. On the opposite fringe of the same cloud, and beyond it, small patches of milky mist would appear, and curdle up, so as to form little cloudlets as dense apparently as the large mass beside which they were formed. The counter processes of production and consumption were evidently

going on at opposite sides of the cloud. Even in
the midst of the serene firmament, where a moment
previously the space seemed absolutely void, white
cloud-patches were formed, their sudden appearance
exciting that kind of surprise which might be
supposed to accompany the observation of a direct
creative act.

These clouds were really the indicators of what
was going on in the unseen air. Without them no
motion was visible; but their appearance and disappearance proved not only the existence of motion,
but also the want of homogeneity in the atmosphere. Though we did not see them, currents
were mingling, possessing different temperatures and
carrying different loads of invisible watery vapour.
We know that clouds are not true vapour, but
vapour precipitated by cold to water. We know
also that the amount of water which the air can
hold in the invisible state depends upon its temperature; the higher the temperature of the air, the
more water will it be able to take up. But, when
a portion of warm air, carrying its invisible charge,
is invaded by a current of low temperature, the
chilled vapour is precipitated, and a cloud is the
consequence. In this way two parcels of moist air,
each of which taken singly may be perfectly transparent, can produce by their mixture an opaque
cloud. In the same way a body of clear humid air,

when it strikes the cold summit of a mountain, may render that mountain 'cloud-capped.'

An illustration of this process, which occurred some years ago in a Swedish ball-room, is recounted by Professor Dove. The weather was clear and cold, and the ball-room was clear and warm. A lady fainted, and air was thought necessary to her restoration. A military officer present tried to open the window, but it was frozen fast. He broke the window with his sword, the cold air entered, and *it snowed in the room.* A minute before this all was clear, the warm air sustaining a large amount of moisture in a transparent condition. When the colder air entered, the vapour was first condensed and then frozen. The admission of cool air even into our London ball-rooms produces mistiness. Mountain-chains are very effective in precipitating the vapour of our south-westerly winds; and this sometimes to such an extent as to produce totally different climates on the two sides of the same mountain-group. This is very strikingly illustrated by the observations of Dr. Lloyd on the rainfall of Ireland. Stations situated on the south-west side of a mountain-range showed a quantity of rain far in excess of that observed upon the north-east side. The winds in passing over the mountains were drained of their moisture, and were afterwards comparatively dry.

Two or three years ago I had an opportunity of witnessing a singular case of condensation at Mortain in Normandy. The tourist will perhaps remember a little chapel perched upon the highest summit in the neighbourhood. A friend and I chanced to be at this point near the hour of sunset. The air was cloudless, and the sun flooded the hillsides and valleys with golden light. We watched him as he gradually approached the crest of a hill, behind which he finally disappeared. Up to this point a sunny landscape of exquisite beauty was spread before us, the atmosphere being very transparent; but now the air seemed suddenly to curdle into mist. Five minutes after the sun had departed, a dense fog filled the valleys and drifted in fleecy masses up the sides of the hills. In an incredibly short time we found ourselves enveloped in local clouds so dense as to render our retreat a matter of some difficulty.

In this case, before the sun had disappeared the air was evidently nearly saturated with transparent vapour. But why did the vapour curdle up so suddenly when the sun departed? Was it because the withdrawal of his beams rendered the *air* of the valleys colder, and thus caused the precipitation of the moisture diffused through the air? No. We must look for an explanation to a more direct action of the sun upon the atmospheric moisture.

Let me explain. The beams which reach us from the sun are of a very composite character. A sheaf of white sunbeams is composed of an infinitude of coloured rays, the resultant effect of all upon the eye being the impression of whiteness. But though the colours, and shades of colour, which enter into the composition of a sunbeam are infinite, for the sake of convenience we divide them into seven, which are known as the prismatic colours.

The beams of the sun, however, produce *heat* as well as light, and there are different *qualities* of heat in the sunbeam as well as different qualities of light—nay, there are copious rays of heat in a sunbeam which give no light at all, some of which never even reach the retina at all, but are totally absorbed by the humours of the eye. Now, the same substance may permit rays of heat of a certain quality to pass freely through it, while it may effectually stop rays of heat of another quality. But in all cases the heat stopped is expended in heating the body which stops it. Now, water possesses this *selecting power* in an eminent degree. It allows the blue rays of the solar beam to pass through it with facility, but it slightly intercepts the red rays, and absorbs with exceeding energy the *obscure rays*; and those are the precise rays which possess the most intense heating power.

We see here at once the powerful antagonism

of the sun to the formation of visible fog, and we
see, also, how the withdrawal of his beams may be
followed by sudden condensation, even before the
air has had any time to cool. As long as the solar
beams swept through the valleys of Mortain, every
particle of water that came in their way was reduced to transparent vapour by the heat which the
particle itself absorbed; or, to speak more strictly,
in the presence of this antagonism precipitation
could not at all occur, and the atmosphere remained
consequently clear.[1] But the moment the sun with-
drew, the vapour followed, without opposition, its
own tendency to condense, and its sudden curdling
up was the consequence.

With regard to the *air*, its temperature may not
only have remained sensibly unchanged for some
time after the setting of the sun, but it may have
actually become warmer through the heat set free
by the act of condensation. It was not, therefore,
the action of cold air upon the vapour which produced the effect, but it was the withdrawal of that
solar energy which water has the power to absorb,
and by absorbing to become dissipated in true
vapour.

I once stood with a friend upon a mountain
which commands a view of the glacier of the Rhone

[1] At this time I was brooding over experiments on the absorption
of radiant heat by aqueous vapour.

from its origin to its end. The day had been one of cloudless splendour, and there was something awful in the darkness of the firmament. This deepening of the blue is believed by those who know the mountains to be an indication of a humid atmosphere. The transparency, however, was wonderful. The summits of Mont Cervin and the Weisshorn stood out in clear definition, while the mighty mass of the Finsteraarhorn rose with perfect sharpness of outline close at hand. As long as the sun was high there was no trace of fog in the valleys, but as he sloped to the west the shadow of the Finsteraarhorn crept over the snow-fields at its base. A dim sea of fog began to form, which after a time rose to a considerable height, and then rolled down like a river along the flanks of the mountain. On entering the valley of the Rhone, it crossed a precipitous barrier, down which it poured like a cataract; but long before it reached the bottom it escaped from the shadow in which it had been engendered, and was hit once more by the direct beams of the sun. Its utter dissipation was the consequence, and though the billows of fog rolled on incessantly from behind, the cloud-river made no progress, but disappeared, as if by magic, where the sunbeams played upon it. The conditions were analogous to those which hold in the case of a glacier. Here the ice-river is incessantly nourished by the mountain snow: it moves down its valley, but does not advance in front. At a certain point the consumption by melting is equal to the supply, and here the glacier ceases. In the case before us the cloud-river, nourished by the incessant condensation of the atmospheric vapour, moved down its valley, but ceased at the point where the dissipating action of the sunbeams equalled the supply from the cloud-generator behind.

VI.

KILLARNEY.

THE total amount of heat which the sun sends annually to the earth is invariable, and hence if any portion of the earth's surface during any given year be colder than ordinary, we may infer with certainty that some other portion of the surface is then warmer than ordinary. The port of Odessa owes its importance to a case of atmospheric compensation of this kind. Forty or fifty years ago, Western Europe received less than its normal amount of heat; the missing sunbeams fell upon the East, and Odessa became, to some extent, the granary from which the hungry West was fed. The position it then assumed it has since maintained. The atmosphere is the grand distributor of heat. It has its cold and warm currents—vast aërial rivers, which chill or cheer according to the proximate sources from which they are derived. In this present year 1860 the British Isles appear to lie near the common boundary of two such currents —the limit, however, shifting so as to cause both to pass over us in swift succession. Near this boundary line the atmospheric currents mingle, and the copious aqueous precipitation which we now observe is the result.

Superadded to this source of general rain, we have at Killarney local condensers in the neighbouring mountains. Round the cool crests of Carrantual and his peaked and craggy brothers the moist and tilted south-west wind curdles ceaselessly into clouds, which nourish the moss and heather whose decomposition produces the peat which clothes the disintegrated rocks. Grandly the vast cumuli build themselves in the atmosphere, hanging at times lazily over the mountains and mottling with their shadows the brown sides of the hills. Reddened by the evening sun, these clouds cast their hues upon the lakes, the crisped surface of which breaks up their images into broad spaces of diffused crimson light. On other days the cumuli seem whipped into dust, and scattered through the general air, mixing therewith as the smoke of London mingles with the supernatant atmosphere. Day by day the guides prophesy fine weather—the blackest cloud is 'all for hate.' You are assured that if you start to-day you will not get 'a single dhrop' of rain ; you go, and are drenched ; but the guide's purpose is accomplished, the moderate sum of three and sixpence being added to his private store.

In ages past these mountain condensers acted differently. The wet winds of the ocean, which now descend in liquid showers upon the hills, once discharged their contents as snow. And a famous deposit they must have made. In addition to the charms which this region presents to every eye, the mind of him who can read the rocks aright is carried back to a time when deep snowbeds cumbered the mountain-slopes, and vast glaciers filled the vales. In neither England nor Wales do the traces of glacial action reach the magnitude which they exhibit here.

The Gap of Dunloe is the channel of an ancient glacier; and all through it the scratching and polishing may be traced. The flanks of the Purple Mountain have been planed down by the moving ice, and the rocky amphitheatre which the guides choose for the production of echoes has been scooped and polished by the same agency. Near the point where the road from the Gap joins that up the Black Valley is a slab of rock, which rivals the famous *Höllen Platte* in Haslithal. The Black Valley, indeed, was the mould through which a great glacier from the adjacent mountains moved, 'unhasting, unresting,' grinding the rocks right and left, and filling the entire basin now occupied by the waters of the Upper Lake. All the islands of this lake are glacier domes. The shapes, moreover, which

have suggested the fanciful names given to some
of the rocks are entirely due to the planing of the
ice. The 'Cannon Rock,' the 'Giant's Coffin,' the
'Man-of-War,' and others, owe their forms to the
mighty moulding-plane which in bygone ages passed
over them.

I have spoken of the echoes in the Gap of Dunloe.
They are very fine, and are usually awakened by a
guide who plays a bugle, and to whom extra wages
are paid on this account. The man times his
operations so that the echo and the original sound
shall not overlap, and he usually places his guests
behind a hill-brow, which partially cuts away the
direct sound, but offers no impediment to the echoes.
He flourishes his trumpet, and pauses; the rocks
respond, the first return of the sound being almost
as strong as the blast itself; the sonorous pulses leap
from crag to crag, and from them to the listener's
ear, diminishing in intensity and augmenting in
softness the oftener they are reflected. Moore's
melody of 'The Meeting of the Waters,' suitably
played, is thus returned with exquisite sweetness by
the reflecting rocks.

The rain here is pitiless, but the march of the
showering clouds over the mountains is sometimes
very grand. One really good day is all that I have
been able to number out of six spent on the banks
of the Lower Lake, and even that day was ushered

in by heavy rain.   Afterwards, however, the cloud
field broke, and the condensed vapours rolled themselves up into sphered masses, which sailed majesti-
cally through the ether.   With some other visitors
I rowed to the Upper Lake, landed at the base of
the Purple Mountain, and with one companion
climbed the latter to its crest.   This is covered by
loose masses of stone of a purplish hue, from which
the mountain derives its name.

A few days previously I had been on the top of
Mangerton, a spot selected by the guides as afford-
ing a prospect of the entire region of the Lakes.
But Mangerton is a stupid mountain, and it is
climbed by a wearisome pony track.   It is incomparably inferior to the Purple Mountain.   From
the latter, on one side, we look into the heart of
Magillicuddy's Reeks, and shake hands with Carrantual across the Gap of Dunloe.   It commands a
splendid mountain panorama, and on the occasion
of my visit showed the Reeks in their true character,
as cloud-generators.   A light wind swept across
them.   Far to westward, towards the sea, the air
was cloudless; but over the Reeks its moisture was
densely precipitated, and formed there a canopy
which threw an inky gloom upon the mountains.
The clouds sometimes descended so as to touch
the summits, but for the most part they floated a
little way above them, leaving the jagged outlines

clear. From the Reeks the clouds were wafted
westward ; but here, meeting with warmer air, they
diminished in size, the smaller ones melting quite
away. Below us gleamed the Upper Lake, running
in and out amid the mountains, fringed with woods
and studded with islands covered with sunny foliage.
From this lake a long, sinuous, and narrow outlet,
called the Long Range, runs to the Middle Lake.
The suddenness with which this lovely sheet of
water opens on quitting the Long Range constitutes
perhaps the greatest surprise which the traveller
here encounters.

We walked along the ridge of the Purple Mountain ankle deep in elastic moss, with glorious views
at either side. Arrived at the end of its greatest
spur, the Middle and Lower Lakes with their
islands, and the wooded and tortuous peninsula
between them, lay before us. No view of the
English lakes known to me could compete in loveliness with this one. We passed onward through the
heather to the brow above the Bay of Glena, and
there clambered down the mountain, helping ourselves by the trees which grasped with gnarled roots
the mossed and slippy crags. At Glena we met
our boat, and were rowed over the jerking waves to
the island of Innisfallen, and thence to our hotel.
Various bits of climbing were accomplished during
my stay, and almost in every case in opposition to

the guides. The Eagle Rock, for example, a truly noble mass, and others, were climbed, amid emphatic enunciations of 'impossible.' Yet these guides and boatmen are fine, hardy fellows, and of great endurance, but they appear averse to trying their strength under new conditions.

I write on a drenching day, and a strong wind which wails dismally round the house has roused the Lower Lake to foam and fury. Innisfallen looms feebly through the grey haze, but the opposite Toumies mountains are plunged in impenetrable gloom. All round the horizon is built a black cloud-wall, but the zenithal heaven is clear. Over the coping of this thunderous bulwark the sun shoots his rays, which, meeting the dropping clouds of the opposite heaven, paint upon it a complete and magnificent bow. Here the white beam enters the front of the falling drop, and is reflected at its back, emerging unravelled to its component hues. But the condition is, that after being thus un-ravelled, the coloured rays shall not diverge on quitting the drop. If they did, they would be lost immediately to the senses; but they are squeezed together to parallel sheaves, and thus their intensity is preserved through long aërial distances. Above the vivid bow hangs its spectral secondary brother, in which a double reflection within each raindrop enfeebles the colours, and inverts the order of succession.

Touched by the wand of law, the dross of facts becomes gold, the meanest being raised thereby to brotherhood with the highest. Thus the smoke of an Irish cabin lifts our speculations to the heavenly dome. We look through the cloudless air at the darkness of infinite space, and are met by the azure of the firmament—we look through a long reach of the same atmosphere at the bright sun or moon and see them orange or red. We look through the peat-smoke at a black rock, or at the dark branches of a yew, and see the smoke blue—we look through the same smoke at a cloud illuminated to whiteness by the sun and find the smoke red. The selfsame column of smoke may be projected against a bright and a dark portion of the same cloud, and thus made to appear blue and red at the same time. The blue belongs to the light *reflected* from the smoke; the red to the light *transmitted* through it. In like manner, the hues of the atmosphere are not due to colouring matter, but to the fact of its being a turbid medium. Through this we look at the blackness of unillumined space and see the blue; at the western heaven at sunset, and meet that light which steeps the clouds of evening in orange and crimson dyes.

VII.

SNOWDON IN WINTER.

TAINTED by the city air, and with gases not natural even to the atmosphere of London, I gladly chimed in with the proposal of an experienced friend to live four clear days at Christmas on Welsh mutton and mountain air. On the evening of the 26th of December 1860 Mr. Busk, Mr. Huxley, and I found ourselves at the Penryhn Arms Hotel in Bangor. Next morning we started betimes. The wind had howled angrily during the night. It now swept over the frozen road, carrying the looser snow along with it, shooting the crystals with projectile force against our faces, and compelling us to lean forward at a considerable angle to keep upon our feet. Our destination was Capel Curig, with a prospective design upon Snowdon ; but we had no bâtons fit for the ascent. At Bethesda, however, after many vain enquiries in Welsh and English about walking-sticks, we found a shop which embraced among its multi-tudinous contents a sheaf of rake-handles. Two of these we purchased at fourpence each, and had them afterwards furnished with rings and iron spikes, at the total cost of one shilling. Thus provided, we hoped that ' old Snowdon's craggy chaos' might be invaded with a hope of success.

On the morning of the 28th we issued from our hotel. A pale blue, dashed with ochre, and blending to a most delicate green, overspread a portion of the eastern sky. Grey cumuli, tinged ruddily here and there as they caught the morning light, swung aloft, but melted more and more as the day advanced. The eastern mountains were all thickly covered with newly-fallen snow. The effect was unspeakably lovely. In front of us was Snowdon; over it and behind it the atmosphere was closely packed with dense brown haze, the lower filaments of which reached almost half-way down the mountain, but still left all its outline clearly visible through the attenuated fog. No ray of sunlight fell upon the hill, and the face which it turned towards us, too steep to hold the snow, exhibited a precipitous slope of rock, faintly tinted by the blue grey of its icy enamel. Below us was Llyn Mymbyr, a frozen plain; behind us the hills were flooded with sunlight, and here and there from the shaded slopes, which were illuminated chiefly by the light of the firmament, shimmered a most delicate blue.

This beautiful effect deserves a word of notice; many doubtless have observed it during the late snow.

Ten days ago, in driving from Kirtlington to Glympton, the window of my cab became partially opaque by the condensation of the vapour of respiration. With the finger-ends little apertures were made in the coating, and when viewed through these the snow-covered landscape flashed incessantly with blue gleams. They rose from the shadows of objects along the road, which shadows were illuminated by the light of the sky. The blue light is best seen when the eye is in motion, thus causing the images of the shadows to pass over different parts of the retina. The whole shadow of a tree may thus be seen with stem and branches of the most delicate blue. I have seen similar effects upon the fresh *névés* of the Alps, the shadow being that of the human body looked at through an aperture in a handkerchief thrown over the face. The same splendid effect was once exhibited in a manner never to be forgotten by those who witnessed it, on the sudden opening of a tent-door at sunrise on the summit of Mont Blanc.

At Pen-y-Gwrid Busk halted, purposing to descend to Llanberis by the road, while Huxley and I went forward to the small public-house known as Pen Pass. Here our guide, Robert Hughes, a powerful but elderly man, refreshed himself, and we quitted the road and proceeded for a short distance along a car-track which seemed to wind round a spur of

Snowdon. 'Is there no shorter way up?' we demanded. 'Yes; but I fear it is now impracticable,'
was the reply. 'Go straight on,' said Huxley, 'and
do not fear us.'

Up the man went with a spurt, suddenly putting
on all his steam. The whisky of Pen Pass had
given him a flash of energy, which we well knew
could not last. In fact, the guide, though he
acquitted himself admirably during the day, had
at first no notion that we should reach the summit;
and this made him careless of preserving himself
at the outset. Toning him down a little, we went
forward at a calmer pace. Crossing the spur, we
came upon a pony-track on the opposite side. It
was rendered conspicuous by the unbroken layer of
snow which rested on it. Huxley took the lead,
wading knee-deep for nearly an hour.

I, wishing to escape this labour, climbed the
slopes to the right, and sought a way over the
less loaded bosses of the mountain. On our remarking to Hughes that he had never assailed
Snowdon under such conditions, he replied that he
had, and under worse. The 12th of April last, he
affirmed, was a worse day, and he had led a lady
on that day almost to the summit. Unluckily for
him, there was a smack of 'bounce' in the reply.
It caused us to conclude that the same energy which
had led the lady could lead us, and hence, when

Huxley fell back, the guide was sent to the front, to break the way. He did this manfully for nearly an hour, at the end of which he seemed very jaded, and as he sat resting on a corner of rock I asked him whether he was tired. 'I am,' was his reply. Huxley gave him a sip of brandy, and I came for a short time to the front.

I had no gaiters, and my boots were incessantly filled with snow. My own heat sufficed for a time to melt the snow; but this clearly could not go on for ever. My left heel first became numbed and painful; and this increased till both feet were in great distress. I sought relief by quitting the track and trying to get along the impending shingle to the right. The high ridges afforded me some relief, but they were separated by couloirs in which the snow had accumulated, and through which I sometimes floundered waist-deep. The pain at length became unbearable; I sat down, took off my boots and emptied them; put them on again; tied Huxley's pocket handkerchief round one ankle, and my own round the other, and went forward once more. It was a great improvement—the pain vanished, and did not return.

The scene was grand in the extreme. Before us were the buttresses of Snowdon, crowned by his conical peak; while below us were three llyns, black as ink, and contracting additional gloom from the shadow of

the mountain.   The lines of weathering had caused
the frozen rime to deposit itself upon the rocks, as
on the tendrils of a vine, the crags being fantasti-
cally wreathed with runners of ice.   The summit,
when we looked at it, damped our ardour a little ;
it seemed very distant, and the day was sinking fast.
From the summit the mountain sloped downward
to a col which linked it with a bold eminence to
our right.   At the col we aimed, and half an hour
before reaching it we passed the steepest portion of
the track.   This I quitted, seeking to cut off the zig-
zags, but gained nothing but trouble by the attempt.
This difficulty conquered, the col was clearly within
reach ; on its curve we met a fine snow cornice,
through which we broke at a plunge, and gained
safe footing on the mountain-rim.   The health and
gladness of that moment were a full recompense
for the entire journey into Wales.

We went upward along the edge of the cone
with the noble sweep of the snow cornice at our
left.   The huts at the top were all cased in ice,
and from their chimneys and projections the snow
was drawn into a kind of plumage by the wind.
The crystals had set themselves so as to present the
exact appearance of feathers, and in some cases
these were stuck against a common axis, so as accurately to resemble the plumes in soldiers' caps.

It was 3 o'clock when we gained the summit. Above and behind us the heavens were of the densest grey; towards the western horizon this was broken by belts of fiery red, which nearer the sun brightened to orange and yellow. The mountains of Flintshire were flooded with glory, and later on, through the gaps in the ranges, the sunlight was poured in coloured beams, which could be tracked through the air to the places on which their radiance fell. The scene would bear comparison with the splendours of the Alps themselves.

Next day we ascended the pass of Llanberis. The waterfalls, stiffened into pillars of blue ice, gave it a grandeur which it might not otherwise exhibit. The wind, moreover, was violent, and shook clouds of snow-dust from the mountain-heads. We descended from Pen-y-Gwrid to Beddgelert. What splendid skating surfaces the lakes presented—so smooth as scarcely to distort the images of the hills! A snow-storm caught us before we reached our hotel. This melted to rain during the night. Next day we engaged a carriage for Carnarvon, but had not proceeded more than two miles when we were stopped by the snow. Huge barriers of it were drifted across the road; and not until the impossibility of the thing was clearly demonstrated did we allow the postilion to back out of his engagement. Luckily our luggage was portable. Strapping our bags and knapsacks on our shoulders, partly through the fields, and partly along the less encumbered portions of the road, we reached Carnarvon on foot, and the evening of the 31st of December saw us safe in London.

VIII.

VOYAGE TO ALGERIA TO OBSERVE THE ECLIPSE.

THE opening of the Eclipse Expedition was not propitious. Portsmouth, on the 5th of December 1870, was swathed by a fog, which was intensified by smoke, and traversed by a drizzle of fine rain. At six P.M. I was on board the 'Urgent.' On Tuesday morning the weather was too thick to permit of the ship's being swung and her compasses calibrated. The Admiral of the port, a man of very noble presence, came on board. Under his stimulus the energy which the weather had damped appeared to become more active, and soon after his departure we steamed down to Spithead. Here the fog had so far lightened as to enable the officers to swing the ship.

At three P.M. on Tuesday the 6th of December we got away, gliding successively past Whitecliff Bay, Bembridge, Sandown, Shanklin, Ventnor, and St. Catherine's Lighthouse. On Wednesday morning we sighted the Isle of Ushant, on the French side of the Channel. The northern end of the island has been fretted by the waves into detached tower-like

masses of rock of very remarkable appearance. In
the Channel the sea was green, and opposite Ushant
it was a brighter green. On Wednesday evening we
committed ourselves to the Bay of Biscay. The roll
of the Atlantic was full, but not violent. There had
been scarcely a gleam of sunshine throughout the
day, but the cloud-forms were fine, and their apparent
solidity impressive. On Thursday morning I rose
refreshed, and found the green of the sea displaced
by a deep indigo blue. The whole of Thursday we
steamed across the bay. We had little blue sky, but
the clouds were again grand and varied—cirrus,
stratus, cumulus, and nimbus, we had them all.
Dusky hairlike trails were sometimes dropped from
the distant clouds to the sea. These were falling
showers, and they sometimes occupied the whole
horizon, while we steamed across the rainless circle
which was thus surrounded. Sometimes we plunged
into the rain, and once or twice, by slightly changing
course, avoided a heavy shower. From time to time
perfect rainbows spanned the heavens from side to side.
At times a bow would appear in fragments, showing
the keystone of the arch midway in air, and its two
buttresses on the horizon. In all cases the light of the
bow could be quenched by a Nicol's prism, with its
long diagonal tangent to the arc. Sometimes gleam-
ing patches of the firmament were seen amid the
clouds. When viewed in the proper direction, the

gleam could be quenched by a Nicol prism, a dark aperture being thus opened into stellar space.

At sunset on Thursday the denser clouds were fiercely fringed, while through the lighter ones seemed to issue the glow of a conflagration. On Friday morning we sighted Cape Finisterre, the extreme end of the arc which sweeps from Ushant round the Bay of Biscay. Calm spaces of blue, in which floated quietly scraps of cumuli, were behind us, but in front of us was a horizon of portentous darkness. It continued thus threatening throughout the day. Towards evening the wind strengthened to a gale, and at dinner it was difficult to preserve the plates and dishes from destruction. Our thinned company hinted that the rolling had other consequences. It was very wild when we went to bed. I slumbered and slept, but after some time was rendered actively conscious that my body had become a kind of projectile, which had the ship's side for a target. I gripped the edge of my berth to save myself from being thrown out. Outside, I could hear somebody say that he had been thrown from his berth, and sent spinning to the other side of the saloon. The screw laboured violently amid the lurching; it incessantly quitted the water, and, twirling in the air, rattled against its bearings, and caused the ship to shudder from stem to stern. At times the waves struck us, not with the soft impact

which might be expected from a liquid, but with the
sudden solid shock of battering-rams. 'No man
knows the force of water,' said one of the officers,
'until he has experienced a storm at sea.' These
blows followed each other at quicker intervals, the
screw rattling after each of them, until, finally, the
delivery of a heavier stroke than ordinary seemed
to reduce the saloon to chaos. Furniture crashed,
glasses rang, and alarmed enquiries immediately followed. Amid the noises I heard one note of forced
laughter; it sounded very ghastly. Men tramped
through the saloon, and busy voices were heard aft,
as if something there had gone wrong.

I rose, and not without difficulty got into my
clothes. In the after-cabin, under the superintendence of the able and energetic navigating lieutenant,
Mr. Brown, a group of blue-jackets were working at
the tiller-ropes. These had become loose, and the
helm refused to answer the wheel. High moral
lessons might be gained on shipboard by observing
what steadfast adherence to an object can accomplish,
and what large effects are heaped up by the addition
of infinitesimals. The tiller-rope, as the blue-jackets
strained in concert, seemed hardly to move; still it
did move a little, until finally, by timing the pull to
the lurching of the ship, the mastery of the rudder
was obtained. I had previously gone on deck.
Round the saloon-door were a few members of the

eclipse party, who seemed in no mood for scientific observation. Nor did I; but I wished to see the storm. I climbed the steps to the poop, exchanged a word with Captain Toynbee, the only member of the party to be seen on the poop, and by his direction made towards a cleat not far from the wheel.[1] Round it I coiled my arms. With the exception of the men at the wheel, who stood as silent as corpses, I was alone.

I had seen grandeur elsewhere, but this was a new form of grandeur to me. The 'Urgent' is long and narrow, and during our expedition she lacked the steadying influence of sufficient ballast. She was for a time practically rudderless, and lay in the trough of the sea. I could see the long ridges, with some hundreds of feet between their crests, rolling upon the ship perfectly parallel to her sides. As they approached they so grew upon the eye as to render the expression 'mountains high' intelligible. At all events, there was no mistaking their mechanical might, as they took the ship upon their shoulders, and swung her like a pendulum. The poop sloped sometimes at an angle which I estimated at over forty-five degrees; wanting my previous Alpine practice, I should have felt less confidence in my grip of the cleat. Here and there the long rollers

[1] The cleat is a T-shaped mass of metal employed for the fastening of ropes.

were tossed by interference into heaps of greater
height. The wind caught their crests, and scattered
them over the sea, the whole surface of which was
seething white. The aspect of the clouds was a fit
accompaniment to the fury of the ocean. The moon
was almost full—at times concealed, at times
revealed, as the scud flew wildly over it. These
things appealed to the eye, while the ear was filled
by the whistle and boom of the storm and the
groaning of the screw.

  Nor was the outward agitation the only object of
interest to me. I was at once subject and object
to myself, and watched with intense interest the
workings of my own mind. The 'Urgent' is an
elderly ship. She had been built, I was told, by a
contracting firm for some foreign Government, and
had been diverted from her first purpose when converted into a troop-ship. She had been for some
time out of work, and I had heard that one of her
boilers, at least, needed repair. Our scanty but
excellent crew, moreover, did not belong to the
'Urgent,' but had been gathered from other ships.
Our three lieutenants were also volunteers. All this
passed swiftly through my mind as the ship shook
under the blows of the waves, and I thought that
probably no one on board could say how much of this
thumping and straining the 'Urgent' would be able
to bear. This uncertainty caused me to look

steadily at the worst, and I tried to strengthen myself in the face of it.

But at length the helm laid hold of the water, and the ship was got gradually round to face the waves. The rolling diminished, a certain amount of pitching taking its place. Our speed had fallen from eleven knots to two. I went again to bed. After a space of calm, when we seemed crossing the vortex of a storm, heavy tossing recommenced. I was afraid to allow myself to fall asleep, as my berth was high, and to be pitched out of it might be attended with bruises, if not with fractures. From Friday at noon to Saturday at noon we accomplished sixty-six miles, or an average of less than three miles an hour. I overheard the sailors talking about this storm. The ' Urgent,' according to those that knew her, had never previously experienced anything like it.[1]

All through Saturday the wind, though somewhat sobered, blew dead against us. The atmospheric effects were exceedingly fine. The cumuli resembled mountains in shape, and their peaked summits shone as white as Alpine snows. At one place this resemblance was greatly strengthened by a vast area of cloud, uniformly illuminated, and lying like a *névé*

---

[1] There is, it will be seen, a fair agreement between these impressions and those so vigorously described by a scientific correspondent of the ' Times.'

below the peaks. From it fell a kind of cloud-river strikingly like a glacier. The horizon at sunset was remarkable—spaces of brilliant green between clouds of fiery red. Rainbows had been frequent throughout the day, and at night a perfectly conti-nuous lunar bow spanned the heavens from side to side. Its colours were feeble; but, contrasted with the black ground against which it rested, its luminousness was extraordinary.

Sunday morning found us opposite to Lisbon, and at midnight we rounded Cape St. Vincent, where the lurching seemed disposed to recommence. Through the kindness of Lieutenant Walton, a cot had been slung for me. It hung between a tiller-wheel and a flue, and at one A.M. I was roused by the banging of the cot against its boundaries. But the wind was now behind us, and we went along at a speed of eleven knots. We felt certain of reach-ing Cadiz by three. But a new lighthouse came in sight, which some affirmed to be Cadiz lighthouse, while the surrounding houses were declared to be Cadiz itself. Out of deference to these statements, the navigating lieutenant changed his course, and steered for the place. A pilot came on board, and he informed us that we were before the mouth of the Guadalquivir, and that the lighthouse was that of Cipiòna. Cadiz was still some eighteen miles distant.

We steered towards the city, hoping to get into

the harbour before dark. But the pilot was snapped
up by another vessel, and we did not get in. We
beat about during the night, and in the morning
found ourselves about fifteen miles from Cadiz. The
sun rose behind the city, and we steered straight into
the light. The three-towered cathedral stood in the
midst, round which swarmed apparently a multitude
of chimney-stacks. A nearer approach showed the
chimneys to be small turrets. A pilot was taken on
board; for there is a dangerous shoal in the harbour.
The appearance of the town as the sun shone upon its
white and lofty walls was singularly beautiful. We
cast anchor; some officials arrived and demanded a
clean bill of health. We had none. They would
have nothing to do with us; so the yellow quaran-
tine flag was hoisted, and we waited for permission to land the Cadiz party. After some hours
of delay the English consul and vice-consul came on
board, and with them a Spanish officer, ablaze with
gold lace and decorations. Under slight pressure
the requisite permission had been granted. We
landed our party, and in the afternoon weighed
anchor. Thanks to the kindness of our excellent pay-
master, I was here transferred to a roomier berth.

Cadiz soon sank beneath the sea, and we sighted
in succession Cape Trafalgar, Tarifa, and the re-
volving light of Ceuta. The water was very calm,
and the moon rose in a quiet heaven. She swung

with her convex surface downwards, the common
boundary between light and shadow being almost
horizontal. A pillar of reflected light shimmered
up to us from the slightly rippled sea. I had already
noticed the phosphorescence of the water, but to-
night it was stronger than usual, especially among
the foam at the bows. A bucket let down into the
sea brought up a number of the little sparkling
organisms which cause the phosphorescence. I
caught some of them in my hand. And here an
appearance was observed which was new to most of
us, and strikingly beautiful to all. Standing at the
bow and looking forwards, at a distance of forty or
fifty yards from the ship a number of luminous
streamers were seen rushing towards us. On nearing
the vessel they rapidly turned, like a comet round
its perihelion, placed themselves side by side, and,
as parallel trails of light, kept up with the ship. One
of them placed itself right in front of the bow as a
pioneer. These comets of the sea were joined at
intervals by others. Sometimes as many as six at a
time would rush at us, bend with extraordinary
rapidity round a sharp curve, and afterwards keep us
company. Leaning over the bow, and scanning the
streamers closely, the frontal portion of each revealed
the outline of a porpoise. The rush of the creatures
through the water had started the phosphorescence,
every spark of which was converted by the motion

of the retina into a line of light. Each porpoise
was thus wrapped in a luminous sheath. The phosphorescence did not cease at the creature's tail, but
was carried many porpoise-lengths behind it.

To our right we had the African hills, illuminated
by the moon. Gibraltar Rock at length became
visible, but the town remained long hidden by a belt
of haze. Through this at length the brighter lamps
struggled. It was like the gradual resolution of a
nebula into stars. As the intervening depth became gradually less the mist vanished more and
more, and finally all the lamps shone through it.
They formed a bright foil to the sombre mass of rock
above them. The sea was so calm and the scene so
lovely that Mr. Huggins and myself stayed on deck
till the ship was moored, near midnight. During
our walking to and fro a striking enlargement of
the disc of Jupiter was observed whenever the heated
air of the funnels came between us and the planet.
On passing away from the heated air, the flat dim
disc would immediately shrink to a luminous point.
The effect was one of retinal persistence. The retinal
image of the planet was set quivering in all azimuths
by the streams of heated air, describing in quick
succession minute lines of light, which summed
themselves to a disc of sensible area.

At six o'clock next morning the gun at the signal
station on the summit of the rock boomed. At eight

the band on board the 'Trafalgar' training-ship,
which was in the harbour, struck up the national
anthem; and immediately afterwards a crowd of
mite-like cadets swarmed up the rigging   After the
removal of the apparatus belonging to the Gibraltar party we went on shore. Winter was in England when we left, but here we had the warmth
of summer. The vegetation was luxuriant—palm-
trees, cactuses, and aloes, all ablaze with scarlet
flowers. A visit to the Governor was proposed, as
an act of necessary courtesy, and I accompanied Admiral Ommaney and Mr. Huggins to the Convent, or
Government House. We sent in our cards, waited
for a time, and were then conducted by an orderly
to his Excellency. He is a fine old man, over six
foot high, and of frank military bearing. He received us and conversed with us in a very genial
manner. He took us to see his garden, his palms,
his shaded promenades, and his orange-trees loaded
with fruit, in all of which he took manifest delight. Evidently 'the hero of Kars' had fallen
upon quarters after his own heart. He appeared full
of good nature, and engaged us on the spot to dine
with him that day.

We sought the town-major for a pass to visit the
lines. While awaiting his arrival I purchased a
stock of white glass bottles, with a view to experiments on the colour of the sea. Mr. Huggins and

myself, who wished to see the rock, were taken by
Captain Salmond to the library, where a model of
Gibraltar is kept, and where we had a capital preliminary lesson.   At the library we met Colonel
Maberly, a courteous and kindly man, who gave us
good advice regarding our excursion.   He sent an
orderly with us to the entrance of the lines.   The
orderly handed us over to an intelligent Irishman,
who was directed to show us everything that we
desired to see, and to hide nothing from us.   We
took the ' upper lines,' traversed the galleries hewn
through the limestone, looked through the em-
brasures, which opened like doors in the precipice,
over the hills of Spain, reached St. George's Hall, and
went still higher, emerging on the summit of one
of the noblest cliffs I have ever seen.

Beyond were the Spanish lines, marked by a line
of white sentry-boxes ; nearer were the English lines,
less conspicuously marked out; and between both
was the neutral ground.   Behind the Spanish lines
was the conical hill called the Queen of Spain's
Chair.   The general aspect of Spain from the rock
is bold and rugged.   Doubling back from the galleries, we struck upwards towards the crest, reached
the signal station, where we indulged in shandy-gaff
and bread and cheese.   Thence to O'Hara's Tower,
the highest point of the rock.   It was built by a
former Governor, who, forgetful of the laws of

terrestrial curvature, thought he might look from the tower into the port of Cadiz. The tower is riven, and may be climbed along the edges of the crack. We got to the top of it; thence descended the curious Mediterranean Stair—a zigzag, mostly of steps down a steeply falling slope, amid palmetto brush, aloes, and prickly pear.

Passing over the Windmill Hill, we were joined at the ' Governor's Cottage' by a car, and drove afterwards to the lighthouse at Europa Point. The tower was built, I believe, by Queen Adelaide, and it contains a fine dioptric apparatus of the first order, constructed by the Messrs. Chance of Birmingham. At the appointed hour we were at the Convent. During dinner the same genial traits which appeared in the morning were still more conspicuous. The freshness of the Governor's nature showed itself best when he spoke of his old antagonist in arms, Mouravieff. Chivalry in war is consistent with its stern prosecution. These two men were chivalrous, and after striking the last blow became friends for ever. Our kind and courteous reception at Gibraltar is a thing to be remembered with pleasure.

On the 15th of December we committed ourselves to the Mediterranean. The views of Gibraltar with which we are most acquainted represent it as a huge ridge; but its aspect, end on, both from the Spanish lines and from the other side, is truly

noble. There is a sloping bank of sand at the back
of the rock, which I was disposed to regard simply
as the *débris* of the limestone. I wished to let
myself down upon it, but had not the time. My
friend Mr. Busk, however, assures me that it is
silica, and that the same sand constitutes the
adjacent neutral ground. There are theories afloat
as to its having been blown from Sahara. The
Mediterranean throughout this first day, and indeed
throughout the entire voyage to Oran, was of less
deep a blue than the Atlantic. Possibly the quantity of organisms may have modified the colour.
At night the phosphorescence was delicious, breaking with the suddenness of a snapped spring along
the crests of the waves formed by the port and
starboard bows. Its strength was not uniform.
Having flashed brilliantly for a time, it would in
part subside, and afterwards regain its vigour.
Several large phosphorescent masses of wierd appearance also floated past.

On the morning of the 16th we sighted the fort
and lighthouse of Marsa el Kibir, and beyond them
the white walls of Oran lying in the bight of a bay,
sheltered by dominant hills. The sun was shining
brightly; during our whole voyage we had not had so
fine a day. The wisdom which had led us to choose
Oran as our place of observation seemed demonstrated. A rather excitable pilot came on board,

and he guided us in behind the Mole, which had suffered much damage last year from an unexplained
outburst of waves from the Mediterranean. Both
port and bow anchors were cast into deep water.
With three huge hawsers the ship's stern was made
fast to three gun-pillars fixed in the Mole ; and here
for a time the ' Urgent' rested from her labours.

M. Janssen, who had rendered his name celebrated
by his observations of the eclipse in India in 1868,
when he showed the solar flames to be eruptions of
incandescent hydrogen, was already encamped in
the open country about eight miles from Oran. On
the 2nd of December he had quitted Paris in a
balloon, with a strong young sailor as his assistant,
had descended near the mouth of the Loire, seen M.
Gambetta, and received from him encouragement and
aid. On the day of our arrival his encampment was
visited by Mr. Huggins, and the kind and courteous
Engineer of the Port drove me subsequently in his
own phaeton to the place. It bore the best repute as
regards freedom from haze and fog, and commanded
an open outlook, but it was inconvenient for us on
account of its distance from the ship. The place
next in repute was the railway station, between
two and three miles distant from the Mole. It was
inspected, but, being enclosed, was abandoned for an
eminence in an adjacent garden, the property of Mr.
Hinshelwood, a Scotchman who had settled some years

previously as an esparto merchant in Oran,[1] and who in the most liberal manner placed his ground at the disposition of the party. Here the tents were pitched on the Saturday by Captain Salmond and his intelligent corps of sappers, the instruments being erected on the Monday under cover of the tents.

Close to the railway station runs a new loopholed wall of defence, through which the highway passes into the open country. Standing on the highway, and looking southwards, about twenty yards to the right is a small bastionet, intended to carry a gun or two. Its roof I thought would form an admirable basis for my telescope, while the view of the surrounding country was unimpeded in all directions. The authorities kindly allowed me the use of this bastionet. Two men, one a blue-jacket named Elliot, and the other a marine named Hill, were placed at my disposal by Lieutenant Walton; and thus aided, on Monday morning I mounted my telescope. The instrument was new to me, and I wished to master all the details of its manipulation.

After some hours of discipline, and as the day was sobering towards twilight, the telescope was dismounted and put under cover. Mr. Huggins joined me, and we visited together the Arab quarter of Oran. The flat-roofed houses appeared very

[1] Esparto is a kind of grass now much used in the manufacture of paper.

clean and white. The street was filled with
loiterers, and the thresholds were occupied by
picturesque groups. Some of the men are very
fine ; we saw many straight, manly fellows who
must have been six foot four in height. They passed
us with perfect indifference, evincing no anger,
suspicion, or curiosity, hardly caring in fact to glance
at us as we passed. In one instance only during my
stay at Oran was I spoken to by an Arab. He was a
tall, good-humoured fellow, who came smiling up
to me, and muttered something about ' les Anglais.'
The mixed population of Oran is picturesque in
the highest degree : the Jews, rich and poor, varying in their costumes as their wealth varies—the
Arabs more picturesque still, and of all shades of
complexion—the negroes, the Spaniards, the French,
all grouped together, and each preserving their
own individuality, formed a picture intensely interesting me.

On Tuesday, the 20th, I was early at the bastionet,
with the view of schooling both myself and my men.
The night had been very squally. The sergeant of
the sappers took charge of our key, and on Tuesday
morning Elliot went for it. He brought back the
intelligence that the tents had been blown down,
and the instruments overturned. Among these was
a large and valuable equatorial from the Royal
Observatory, Greenwich. It seemed hardly possible

that this instrument, with its wheels and verniers and delicate adjustments, could have escaped uninjured from such a fall. This, however, was the case; and during the day all the overturned instruments were restored to their places, and found to be in practical working order. This and the following day were devoted to incessant schooling. I had come out as a general star-gazer, and not with the intention of devoting myself to the observation of any particular phenomenon. I wished to see the whole—the first contact, the advance of the moon, and the successive swallowing up of the solar spots, the breaking of the last line of crescent by the lunar mountains into Bailey's beads, the advance of the shadow through the air, the appearance of the corona and prominences at the moment of totality, the radiant streamers of the corona, the internal structure of the flames, a glance through a polariscope, a sweep round the landscape with the naked eye, the reappearance of the solar limb through Bailey's beads, and, finally, the retreat of the lunar shadow through the air.

For these observations I was provided with a telescope of admirable definition, mounted, adjusted, packed, and most liberally placed at my disposal by Mr. Warren De la Rue. The telescope grasped the whole of the sun, and a considerable portion of the space surrounding it. But it would not take in the probable extreme limits of the corona. For this the 'finder' was suitable; but, instead of it, I had lashed on to the large telescope a light but powerful instrument, constructed by Ross, and lent to me by Mr. Huggins. I was also furnished with an excellent binocular by Mr. Dallmeyer. In fact, no man could have been more efficiently supported than I was. It required a strict parcelling out of the two minutes and some seconds of totality to embrace in them the entire series of observations. These, while the sun remained visible, were to be made with an unsilvered diagonal eyepiece, which reflected but a small fraction of the sun's light, this fraction being still further toned down by a dark glass. At the moment of totality the dark glass was to be removed, and a silver reflector pushed in, so as to get the maximum of light from the corona and prominences. The time of totality was distributed as follows:

1. Observe approach of shadow through the air: totality.
2. Telescope   .   .   .   . 30 seconds.
3. Finder   .   .   .   . 30 seconds.
4. Double image prism   .   . 15 seconds.
5. Naked eye .   .   .   . 10 seconds.
6. Finder or binocular   .   . 20 seconds.
7. Telescope   .   .   .   . 20 seconds.
8. Observe retreat of shadow.

It was proposed to begin and end with the telescope, so that any change in the field of view occurring during the totality might be noticed. Elliot stood beside me, watch in hand, and furnished with a lantern. He called out at the end of each interval, and I moved from telescope to finder, from finder to polariscope, from polariscope to naked eye, from naked eye back to finder, from finder to telescope, abandoning the instrument finally to observe the retreating shadow. All this we went over twenty times, while looking at the actual sun, and keeping him in the middle of the field. It was my object to render the repetition of the lesson so mechanical as to leave no room for flurry, forgetfulness, or excitement. Volition was not to be called upon, nor judgment exercised, but a well-beaten path of routine was to be followed. Had the opportunity occurred, I think the programme would have been strictly carried out.

But the opportunity did not occur. For several days the weather had been ill-natured. We had wind so strong as to render the hawsers at the stern of the 'Urgent' as rigid as iron, and, therefore, to destroy the navigating lieutenant's sleep. We had clouds, a thunder-storm, and some rain. Still the hope was held out that the atmosphere would cleanse itself, and if it did we were promised an air of extraordinary limpidity. Early on the 22nd we were all at our posts. Spaces of blue in the early morning gave us some encouragement, but all depended on the relation of these spaces to the surrounding clouds. Which of them were to grow as the day advanced? The wind was high, and to secure the steadiness of my instrument I was forced to retreat behind a projection of the bastionet, place stones upon its stand, and, further, to avail myself of the shelter of a sail. My practised men fastened the sail at the top, and loaded it with boulders at the bottom. It was tried severely, but it stood firm.

The clouds and blue spaces fought for a time with varying success. The sun was hidden and revealed at intervals, hope oscillating in synchronism with the changes of the sky. At the moment of first contact a dense cloud intervened, but a minute or two afterwards the cloud had passed, and the enchroachment of the black body of the moon was evident upon the solar disc. The moon marched onward, and I saw it at frequent intervals; a large group of spots were approached and swallowed up. Subsequently I caught sight of the lunar limb as it cut through the middle of a large spot. The spot was not to be distinguished from the moon, but rose like a mountain above it. The clouds, when thin, could be seen as grey scud drifting across the black surface of the moon; but they thickened more and more, and made the intervals of clearness scantier. During these moments I watched with an interest bordering upon fascination the march of the silver sickle of the sun across the field of the telescope. It was so sharp and so beautiful. No trace of the lunar limb could be observed beyond the sun's boundary. Here, indeed, it could only be relieved by the corona, which was utterly cut off by the dark glass. The blackness of the moon beyond the sun was, in fact, confounded with the blackness of space.

Beside me was Elliot with the watch and lantern, while Lieutenant Archer, of the Royal Engineers, had the kindness to take charge of my note-book. I mentioned, and he wrote rapidly down, such things as seemed worthy of remembrance. Thus my hands and mind were entirely free; but it was all to no purpose. A patch of sunlight fell and rested upon the landscape some miles away. It was the only illuminated spot within view. But to the north-west there was still a space of blue which might reach us in time. Within seven minutes of totality another small space towards the zenith became very dark. The atmosphere was, as it were, on the brink of a precipice; it was charged with humidity, which required but a slight chill to bring it down in clouds. This was furnished by the withdrawal of the solar beams; the clouds did come down, covering up the space of blue on which our hopes had so long rested. I abandoned the telescope and walked to and fro, like a leopard in its

cage. As the moment of totality approached, the
descent towards darkness was as obvious as a falling
stone. I looked towards a distant ridge, where I
knew the darkness would first appear. At the moment
a fan of beams, issuing from the hidden sun, was
spread out over the southern heavens. These
beams are bars of alternate light and shade, produced in illuminated haze by the shadows of
floating cloudlets of varying density. The beams
are really parallel, but by an effect of perspective
they appear divergent, like a fan, having the sun,
in fact, for their point of intersection. The darkness took possession of the ridge to which I have
referred, lowered upon M. Janssen's observatory,
passed over the southern heavens, blotting out the
beams as if a sponge had been drawn across them.
It then took successive possession of three spaces of
blue sky in the south-eastern atmosphere. I again
looked towards the ridge. A glimmer as of day-
dawn was behind it, and immediately afterwards
the fan of beams which had been for more than two
minutes absent revived. The eclipse of 1870 had
ended, and, as far as the corona was concerned,
we had been defeated.

Even in the heart of the eclipse the darkness was
by no means perfect. Small print could be read.
In fact, the clouds which rendered the day a dark
one, by scattering light into the shadow, rendered

it less intense than it would have been had the atmosphere been without cloud. In the more open spaces I sought for stars, but could find none. There was a lull in the wind before and after totality, but during the totality the wind was strong. I waited for some time on the bastionet, hoping to get a glimpse of the moon on the opposite border of the sun, but in vain. The clouds continued, and some rain fell. The day brightened somewhat afterwards, and, having packed all up, in the sober twilight Mr. Crookes and myself climbed the heights above the fort of Vera Cruz. From this eminence we had a very noble view over the Mediterranean and the flanking African hills. The sunset was remarkable, and the whole outlook exceedingly fine.

The able and well-instructed medical officer of the 'Urgent,' Mr. Goodman, observed the following temperatures during the progress of the eclipse:

Hour	Deg.	Hour	Deg.
11.45	. 56	12.43	. 51
11.55	. 55	1.5	. 52
12.10	. 54	1.27	. 53
12.37	. 53	1.44	. 56
12.39	. 52	2.10	. 57

The minimum temperature occurred some minutes after totality, when a slight rain fell.

The wind was so strong on the 23rd that Captain Henderson would not venture out. Guided by Mr.

Goodman, I visited a cave scooped into a remarkable stratum of shell-breccia, and, thanks to my guide, secured specimens. Mr. Busk informs me that a precisely similar breccia is found at Gibraltar at approximately the same level. During the afternoon Admiral Ommaney and myself drove to the fort of Marsa el Kibir. The fortification is of ancient origin, the Moorish arches being still there in decay, but the fort is now very strong. About four or five hundred dragoons, fine-looking men, were looking after their horses, waiting for a lull to enable them to embark for France. One of their officers was wandering in a very solitary fashion over the fort. We had some conversation with him. He had been at Sedan, had been taken prisoner, but had effected his escape. He shook his head when we spoke of the termination of the war, and predicted its long continuance. There was bitterness in his tone as he spoke of the charges of treason which had been so lightly levelled against French commanders. The green waves raved round the promontory on which the fort stands, smiting the rocks, breaking into snow, and jumping, after impact, to a height of a hundred feet and more into the air. On our return our vehicle broke down through the loss of a wheel. The Admiral went on board, while I hung long over the agitated sea. The little horses of Oran well merit a passing

word. Their speed and endurance, which are both
heavily drawn upon by their drivers, are extraordinary.

The wind sinking, we lifted anchor on the 24th.
For some hours we went pleasantly along; but
during the afternoon the storm revived, and it blew
heavily against us all the night. When we came
opposite the Bay of Almeria, on the 25th, the
captain turned the ship, and steered into the bay,
where, under the shadow of the Sierra Nevada, we
passed Christmas night in peace. Next morning
' a rose of dawn ' rested on the snows of the adjacent
mountains, while a purple haze was spread over the
lower hills. I had no notion that Spain possessed
so fine a range of mountains as the Sierra Nevada.
The height is considerable, but the form also is
such as to get the maximum of grandeur out of the
height. We got away at eight A.M., passing for a
time through shoal water, the bottom of which had
been evidently stirred up. The adjacent land
seemed eroded in a remarkable manner. Doubtless
it has its times of flood, which excavate these
valleys and ravines, and leave those singular ridges
behind. Towards evening I climbed the mainmast,
and, standing on the crosstrees, saw the sun set
amid a blaze of fiery clouds. The wind was strong
and bitterly cold, and I was glad to return to
the deck along a rope which stretched from the

mast-head to the ship's side. That night we cast
anchor beside the Mole of Gibraltar.

On the morning of the 27th, in company with
two friends, I drove to the Spanish lines, with the
view of seeing the rock from that side. It is an
exceedingly noble mass. The Peninsular and Ori-
ental mail-boat had been signalled and had come.
Heavy duties called me homeward, and by trans-
ferring myself from the 'Urgent' to the mail-
steamer I should gain three days. I hired a boat,
rowed to the steamer, learned that she was to start
at one, and returned with all speed to the 'Urgent.'
Making known to Captain Henderson my wish to
get away, he expressed doubts as to the possibility
of reaching the mail-steamer in time. With his
accustomed kindness, he, however, placed a boat at
my disposal. Four hardy fellows and one of the
ship's officers jumped into it; my luggage, hastily
thrown together, was tumbled in afterwards, and we
were immediately on our way. We had nearly four
miles to row in about twenty minutes; but we
hoped the mail-boat might not be punctual. For a
time we watched her anxiously; there was no motion;
we came nearer, but the flags were not yet hauled
in. The men put forth all their strength, animated
by the exhortations of the officer at the helm. The
roughness of the sea rendered their efforts to some
extent nugatory: still we were rapidly approaching

the steamer. At length she moved, punctually
almost to the minute, at first slowly, but soon with
quickened pace. We turned to the left, so as to
cut across her bows. Five minutes' pull would have
brought us up to her. The officer waved his cap
and I my hat. 'If they could only see us, they
might back to us in a moment.' But they did not
see us, or if they did, they paid no attention to us.
I returned to the 'Urgent,' discomfited, but grateful
to the fine fellows who had wrought so hard to carry
out my wishes.

Glad of the quiet, in the sober afternoon I took
a walk towards Europa Point. The sky darkened,
and heavy squalls passed at intervals. Rain began
to fall, and I returned home. Private theatricals
were at the Convent, and the kind and courteous
Governor had sent cards to the eclipse party. I
failed in my duty in not going. I had heard of
St. Michael's Cave as rivalling, if not outrivalling,
the Mammoth Cave of Kentucky. On the 28th
Messrs. Crookes, Carpenter, and myself, guided by a
military policeman who understood his work, ex-
plored the cavern. The mouth is about 1,100 feet
above the sea. We zigzagged up to it, and first
were led into an aperture in the rock some height
above the true entrance of the cave. In this upper
cavern we saw some tall and beautiful stalactite
pillars.

The water drips from the roof charged with bicarbonate of lime. Exposed to the air, the carbonic acid partially escapes, and the simple carbonate of lime, which is hardly at all soluble in water, deposits itself as a solid, forming stalactites and stalagmites. Even the exposure of chalk or limestone water to the open air partially softens it. A specimen of the Redbourne water exposed by Messrs. Graham, Miller, and Hofmann in a shallow basin, fell from eighteen degrees to nine degrees of hardness. The softening process of Clark is virtually a hastening of the natural process. Here, however, instead of being permitted to evaporate, half the carbonic acid is appropriated by lime, the half thus taken up, as well as the remaining half, being precipitated. The solid precipitate is permitted to sink, and the clear supernatant liquid is limpid soft water.

We returned to the real mouth of St. Michael's Cave, which is entered by a wicket. The floor was somewhat muddy, and the roof and walls were wet. Our guide took off his coat, but we did not follow his example. We were soon in the midst of a natural temple, where tall columns sprang complete from floor to roof, while incipient columns were growing to meet each other, upwards and downwards. The water which trickles from the stalactite, after having in part yielded up its carbonate of lime, falls

upon the floor vertically underneath, and there
builds the stalagmite. Consequently, the pillars
grow from above and below simultaneously along
the same vertical. It is easy to distinguish the
stalagmitic from the stalactitic portion of the
pillars. The former is always divided into short
segments by protuberant rings, as if deposited
periodically, while the latter presents a uniform
surface. In some cases the points of inverted cones
of stalactite rested on the centres of pillars of
stalagmite. The process of solidification and the
architecture were alike beautiful.

We followed our guide through various branches
and arms of the cave, climbed and descended steps,
halted at the edges of dark shafts and apertures,
squeezed ourselves through narrow passages, where
the sober grey of my coat suffered less than the black
of my companions'. From time to time we halted,
while Mr. Crookes illuminated with ignited magne-
sium wire the roof, columns, dependent spears, and
graceful drapery of the stalactite. Once, coming to
a magnificent cluster of icicle-like spears, we helped
ourselves to specimens. There was some difficulty
in detaching the more delicate ones, their fragility
was so great. A consciousness of vandalism which
smote me at the time haunts me still; for, though
our requisitions were moderate, this beauty ought
not to be at all invaded. Pendent from the roof in

their natural habitat, nothing can exceed their delicate beauty; they *live*, as it were, surrounded by organic connections. In London they are curious, but not beautiful. Of gathered shells Emerson writes :

> I wiped away the weeds and foam,
> And brought my sea-born treasures home:
> But the poor, unsightly, noisome things
> Had left their beauty on the shore,
> With the sun, and the sand, and the wild uproar.

The promontory of Gibraltar is so burrowed with caverns that it has been called the Hill of Caves. They are apparently related to the geologic disturbances which the rock has undergone. The earliest of these is the tilting of the once horizontal strata. Suppose a force acting upon the promontory at its southern extremity, near Europa Point, tending to twist the strata in a direction opposed to that of the hands of a watch, and suppose the rock to be of a partially yielding character, such a force would turn the strata into screw-surfaces, the greatest amount of twisting being endured near the point of application of the force. Such a twisting the rock appears to have suffered; but instead of the twist fading gradually and uniformly off in passing from south to north, the want of uniformity in the material has produced lines of dislocation where there are abrupt changes in the amount of twist. Thus, at the northern end of the rock the dip to the west is nineteen degrees; in the middle hill it is thirty-eight degrees; in the centre of the south hill, or Sugar Loaf, it is fifty-seven degrees. At the southern extremity of the Sugar Loaf the strata are vertical, while further to the south they actually turn over and dip to the east.

The rock is thus divided into three sections, separated from each other by surfaces of dislocation, where the rock is much wrenched and broken. These places of dislocation are called the Northern and Southern Quebrada, from the Spanish 'Tierra Quebrada,' or broken ground; and it is at these places that the inland caves of Gibraltar are almost exclusively found. Based on the observations of Dr. Falconer and himself, an excellent and most interesting account of these caves, and of the human remains and works of art which they contain, was given by Mr. Busk at the meeting of the Congress of Prehistoric Archæology at Norwich, and afterwards printed in the 'Transactions' of the Congress.[1] Long subsequently to the operation of the twisting force just referred to, the promontory underwent various changes of level. There are sea-terraces and layers of shell-breccia along its flanks, and numerous caves which, unlike the inland one, are the product

---

[1] In this essay Mr. Busk refers to the previous labours of Mr. Smith, of Jordan Hill, to whom we owe most of our knowledge of the geology of the rock.

of marine erosion.   The Apes' Hill, on the African side of the strait, Mr. Busk informs me has undergone similar disturbances.[1]

In the harbour of Gibraltar, on the morning of our departure, I resumed a series of observations on the colour of the sea.   On my way out I had collected a number of specimens, with a view to subsequent examination.   But the bottles were claret bottles, and I could by no means feel sure of their purity.   At Gibraltar, therefore, I purchased fifteen white glass bottles, with ground glass stoppers, and at Cadiz, thanks to the friendly guidance of Mr. Cameron, I secured a dozen more.   These seven-and-twenty bottles were filled with water, taken at different places between Oran and Spithead.

And here let me express my warmest acknowledgments to Captain Henderson, the commander of H.M.S. 'Urgent,' who aided me in my observations in every possible way.   Indeed, my best thanks are due to all the officers for their unfailing courtesy and help.   The captain placed at my disposal his own coxswain, an intelligent fellow named Thorogood, who skilfully attached a cord to each bottle, weighted it with lead, cast it into the sea,

[1] No one can rise from the perusal of Mr. Busk's paper without a feeling of admiration for the principal discoverer and indefatigable explorer of the Gibraltar caves, the late Captain Frederick Brome.

and, after three successive rinsings, filled it under my own eyes. The contact of jugs, buckets, or other vessels was thus avoided, and even the necessity of pouring the water out afterwards through the dirty London air.

The mode of examination applied to these bottles after my return to London is in some sense complementary to that of the microscope, and may I think materially aid enquiries conducted with that instrument. In microscopic examination attention is directed to a small portion of the liquid, the aim being to detect the individual suspended particles. In my case, a large portion of the liquid is illuminated by a powerfully condensed beam, its general condition being revealed through the light scattered by suspended particles. Care is taken to defend the eye from the access of all other light, and, thus defended, it becomes an organ of inconceivable delicacy. Were water of uniform density perfectly free from suspended matter, it would, in my opinion, scatter no light at all. The track of a luminous beam could not, I think, be seen in such water. But an amount of impurity so infinitesimal as to be scarcely expressible in numbers, and the individual particles of which are so small as wholly to elude the microscope, may, when examined by the method alluded to, produce not only sensible, but striking, effects upon the eye.

H H

The results of the examination of nineteen bottles, filled at various places between Gibraltar and Spithead, are here tabulated :

No.	Locality	Colour of Sea	Appearance in Electric Beam
1	Gibraltar Harbour . .	Green . . .	Thick with fine particles
2	Two miles from Gibraltar	Clearer green	Thick with very fine particles
3	Off Cabreta Point . . .	Bright green	Still thick, but less so
4	Off Cabreta Point . . .	Black-indigo	Much less thick, very pure
5	Off Tarifa	Undecided .	Thicker than No. 4
6	Beyond Tarifa	Cobalt-blue .	Much purer than No. 5
7	Twelve miles from Cadiz	Yellow-green	Very thick
8	Cadiz Harbour	Yellow-green	Exceedingly thick
9	Fourteen miles from Cadiz	Yellow-green	Thick, but less so
10	Fourteen miles from Cadiz	Bright green	Much less thick
11	Between Capes St. Mary and Vincent	Deep indigo.	Very little matter, very pure
12	Off the Burlings . . .	Strong green	Thick with fine matter
13	Beyond the Burlings . .	Indigo . .	Very little matter, pure
14	Off Cape Finisterre . .	Undecided .	Less pure
15	Bay of Biscay	Black-indigo	Very little matter, very pure
16	Bay of Biscay	Indigo . .	Very fine matter. Iridescent
17	Off Ushant	Dark green .	A good deal of matter
18	Off St. Catherine's . . .	Yellow-green	Exceedingly thick
19	Spithead	Green . . .	Exceedingly thick

Here, in the first instance, we have three specimens of water, described as green, a clearer green, and bright green, taken in Gibraltar Harbour, at a point two miles from the harbour, and off Cabreta Point. The home examination showed that the first was thick with suspended matter, the second less thick, and the third still less thick. Thus the green brightened as the suspended matter became less.

Previous to the fourth observation our excellent navigating lieutenant, Mr. Brown, steered along the coast, thus avoiding the adverse current which sets in through the Strait of Gibraltar from the

Atlantic to the Mediterranean. He was at length forced to cross the boundary of the Atlantic current, which was defined with extraordinary sharpness. On the one side of it the water was a vivid green, on the other a deep blue. Standing at the bow of the ship, a bottle could be filled with blue water, while at the same moment a bottle cast from the stern could be filled with bright green water. Two bottles were secured, one on each side of this remarkable boundary. In the distance the Atlantic had the hue called ultramarine; but looked fairly down upon, it was of almost inky blackness—black qualified by a trace of indigo.

What change does the home examination here reveal? In passing to indigo, the water becomes suddenly augmented in purity, the suspended matter has become suddenly less. Off Tarifa, the deep indigo disappears, and the sea is undecided in colour. Accompanying this change, we have a rise in the quantity of suspended matter. Beyond Tarifa, we change to cobalt-blue, the suspended matter falling at the same time in quantity. This water is distinctly purer than the green. We approach Cadiz, and at twelve miles from the city get into yellow-green water; this the London examination shows to be thick with suspended matter. The same is true of Cadiz Harbour, and also of a point fourteen miles from Cadiz in the homeward direction. Here there is a sudden change from yellow-green to a bright emerald-green, and accompanying the change a sudden fall in the quantity of suspended matter. Between Cape St. Mary and Cape St. Vincent the water changes to the deepest indigo. In point of purity, this indigo water is shown by the home examination to transcend the emerald-green water.

We now reach the remarkable group of rocks called the Burlings, and find the water between the shore and the rocks a strong green; the home examination shows it to be thick with fine matter. Fifteen or twenty miles beyond the Burlings we come again into indigo water, from which the suspended matter has in great part disappeared. Off Cape Finisterre, about the place where the 'Captain' went down, the water becomes green, and the home examination pronounces it to be thicker. Then we enter the Bay of Biscay, where the indigo resumes its power, and where the home examination shows the greatly augmented purity of the water. A second specimen of water taken from the Bay of Biscay held in suspension fine particles of a peculiar kind; the size of them was such as to render the water richly iridescent. It showed itself green, blue, or salmon colour, according to the direction of the line of vision. Finally, we come to our last two bottles, the one taken opposite St. Catherine's lighthouse, in the Isle of Wight, the other at Spithead. The sea at both these places

was green, and both specimens, as might be expected,
were pronounced by the home examination to be
thick with suspended matter.

Two distinct series of observations are here referred to—the one consisting of direct observations
of the colour of the sea, conducted during the
voyage from Gibraltar to Portsmouth; the other
conducted in the laboratory of the Royal Institution. And here it is to be noted that in the home
examination I never knew what water I had in my
hands. The labels, which had written upon them
the names of the localities, had been tied up, all
information regarding the source of the water being
thus precluded. The bottles were simply numbered,
and not till all the waters had been examined were
the labels opened, and the locality and sea-colour
corresponding to the various specimens ascertained.
I must, therefore, have been perfectly unbiassed in
my home observations, and they, I think, clearly
establish the association of the green colour of sea-
water with fine suspended matter, and the association
of the ultramarine colour, and more especially of
the black-indigo hue of sea-water, with the comparative absence of such matter.

What, in the first place, is the cause of the dark
hue of the deep ocean? [1]   A preliminary remark or

---

[1] A note, written to me on October 22, by my friend Canon
Kingsley, contains the following reference to this point: 'I have

two will clear our way towards an explanation. Colour resides in white light, appearing generally when any constituent of the white light is withdrawn. The hue of a purple liquid, for example, is immediately accounted for by its action on a spectrum. It cuts out the yellow and green, and allows the red and blue to pass through. The blending of these two colours produces the purple. But while the liquid attacks with special energy the yellow and green colours, it enfeebles the whole spectrum; and by increasing the thickness of the stratum we absorb the whole of the light. The colour of a blue liquid is similarly accounted for. It first extinguishes the red; then, as the thickness augments, it attacks the orange, yellow, and green in succession; the blue alone finally remaining. But even it might be extinguished by a sufficient depth of liquid.

And now we are prepared for a brief, but tolerably complete, statement of the action of sea-water upon light, to which it owes its darkness. The spectrum embraces three classes of rays—the thermal, the visual, and the chemical. These divisions overlap each other; the thermal rays are in part visual, the

never seen the Lake of Geneva, but I thought of the brilliant dazzling dark blue of the mid-Atlantic under the sunlight, and its black-blue under cloud, both so solid that one might leap off the sponson on to it without fear; this was to me the most wonderful thing which I saw on my voyages to and from the West Indies.'

visual rays in part chemical, and *vice versâ.* The vast body of thermal rays is beyond the red, being invisible. These rays are attacked with exceeding energy by water. They are absorbed close to the surface of the sea, and are the great agents in evaporation. At the same time the whole spectrum suffers enfeeblement; water attacks all its rays, but with different degrees of energy. Of the visual rays, the red are attacked first, and first extinguished. While the red is disappearing the remaining colours are enfeebled. As the solar beam plunges deeper into the sea, orange follows red, yellow follows orange, green follows yellow, and the various shades of blue, where the water is deep enough, follow green. Absolute extinction of the solar beam would be the consequence if the water were deep and uniform; and if it contained no suspended matter, such water would be as black as ink. A reflected glimmer of ordinary light would reach us from its surface, as it would from the surface of actual ink; but no light, hence no colour, would reach us from the body of the water.

In very clear and very deep sea-water this condition is approximately fulfilled, and hence the extraordinary darkness of such water. The indigo, to which I have already referred, is, I believe, to be ascribed in part to the suspended matter, which is never absent, even in the purest natural water, and in part to the slight reflection of the light from the limiting surfaces of strata of different densities. A modicum of light is thus thrown back to the eye before the depth necessary to absolute extinction has been attained. An effect precisely similar occurs under the moraines of the Swiss glaciers. The ice here is exceptionally compact, and, owing to the absence of the internal scattering common in bubbled ice, the light plunges into the mass, is extinguished, and the perfectly clear ice presents an appearance of pitchy blackness.[1]

The green colour of the sea when it contains matter in a state of mechanical suspension has now to be accounted for, and here, again, let us fall back upon the sure basis of experiment. A strong white dinner-plate was surrounded securely by cord, and had a lead weight fastened to it. Fifty or sixty yards of strong hempen line were attached to the plate. With it in his hand, my assistant, Thorogood, occupied a boat fastened as usual to the davits of the ' Urgent,' while I occupied a second boat nearer to the stern of the ship. He cast the plate as a mariner heaves the lead, and by the time it had reached me it had sunk a considerable depth in the water. In all cases the hue of this plate was green : even when the sea was of the darkest indigo, the

[1] I learn from a correspondent that certain Welsh tarns, which are reputed bottomless, have this inky hue.

green was vivid and pronounced. I could notice the gradual deepening of the colour as the plate sank, but at its greatest depth in indigo water the colour was still a blue-green.[1]

Other observations confirmed this one. The 'Urgent' is a screw steamer, and right over the blades of the screw was an orifice called the screw-well, through which one could look from the poop down upon the screw. The surface glimmer which so pesters the eye was here in a great measure removed. Midway down a plank crossed the screw-well from side to side, and on this I used to place myself to observe the action of the screw underneath. The eye was rendered sensitive by the moderation of the light, and, still further to remove all disturbing causes, Lieutenant Walton had a sail and tarpaulin thrown over the mouth of the well. Underneath this I perched myself and watched the screw. In an indigo sea the play of colour was indescribably beautiful, and the contrast between the water which had the screw-blades for a background, and that which had the bottom of the ocean as a background, was extraordinary. The one was of the most brilliant green, the other of the deepest ultramarine. The surface of the water above the screw-blade was always ruffled. Liquid lenses

---

[1] In no case, of course, is the green pure, but a mixture of green and blue.

were thus formed, by which the coloured light was
withdrawn from some places and concentrated upon
others, the colour being thus caused to flash with
metallic lustre.   The screw-blades in this case
played the part of the plate in the former case,
and there were other instances of a similar kind.
The white bellies of the porpoises showed the green
hue, varying in intensity as the creatures swung to
and fro between the surface and the deeper water.
Foam, at a certain depth below the surface, is also
green.   In a rough sea the light which has penetrated the summit of a wave sometimes reaches the
eye, a beautiful green cap being thus placed upon
the wave even in indigo water.

But how is this colour to be connected philo-
sophically with the suspended particles ?   Take the
dinner-plate which showed so brilliant a green when
thrown into indigo water.   Suppose it to diminish
in size until it reaches an almost microscopic magnitude.   It would still behave substantially as the
larger plate, sending to the eye its modicum of green
light.   If the plate, instead of being a large coherent
mass, were ground to a powder sufficiently fine, and
in this condition diffused through the clear sea-water,
it would send green light to the eye.   In fact, the
suspended particles which the home examination
reveals act in all essential particulars like the plate,
or like the screw-blades, or like the foam, or like

the bellies of the porpoises.   Thus I think the greenness of the sea is physically connected with the matter which it holds in suspension.

We reached Portsmouth on the 5th of January 1871.   There ended a voyage which, though its main object was not realised, has left behind it pleasant memories, both of the aspects of nature and the genial kindliness of men.

LONDON: PRINTED BY
SPOTTISWOODE AND CO., NEW-STREET SQUARE
AND PARLIAMENT STREET

Printed in the United States
By Bookmasters